More than a Glitch

More than a Glitch

Confronting Race, Gender, and Ability Bias in Tech

Meredith Broussard

The MIT Press

Cambridge, Massachusetts | London, England

First MIT Press paperback edition, 2024

An earlier version of chapter 5, Real Students, Imaginary Grades, was published as "When Algorithms Give Real Students Imaginary Grades" in the *New York Times*, September 9, 2020. https://www.nytimes.com/2020/09/08/opinion/international -baccalaureate-algorithm-grades.html.

An earlier version of chapter 7, Gender Rights and Databases, was published as "The Next Frontier for Gender Rights Is Inside Databases," in *"You Are Not Expected to Understand This": How 26 Lines of Code Changed the World*, ed. Torie Bosch (Princeton, NJ: Princeton University Press, 2022). All rights reserved.

The MIT Press would like to thank the anonymous peer reviewers who provided comments on drafts of this book. The generous work of academic experts is essential for establishing the authority and quality of our publications. We acknowledge with gratitude the contributions of these otherwise uncredited readers.

This book was set in Stone Serif and Stone Sans by Westchester Publishing Services. Printed and bound in the United States of America.

Library of Congress Cataloging-in-Publication Data

Names: Broussard, Meredith, author.
Title: More than a glitch : confronting race, gender, and ability bias in tech / Meredith Broussard.
Description: Cambridge, Massachusetts : The MIT Press, [2023] | Includes bibliographical references and index. | Summary: "Broussard argues that the structural inequalities reproduced in algorithmic systems are no glitch. They are part of the system design. This book shows how everyday technologies embody racist, sexist, and ableist ideas; how they produce discriminatory and harmful outcomes; and how this can be challenged and changed"— Provided by publisher.
Identifiers: LCCN 2022019913 (print) | LCCN 2022019914 (ebook) | ISBN 9780262047654 | ISBN 9780262373067 (epub) | ISBN 9780262373050 (pdf) ISBN 9780262548328 (pb.)
Subjects: LCSH: Technology—Social aspects. | Data processing—Social aspects. | Artificial intelligence—Social aspects. | Discrimination. | Software failures.
Classification: LCC T14.5 .B765 2023 (print) | LCC T14.5 (ebook) | DDC 303.48/3—dc23/eng/20221006
LC record available at https://lccn.loc.gov/2022019913
LC ebook record available at https://lccn.loc.gov/2022019914

10 9 8 7 6 5 4

For my family

Contents

1

Introduction

Often, when people talk about making more equitable technology, they start with "fairness." This is a step in the right direction. Unfortunately, it is not a big enough step in the right direction. Understanding why starts with a cookie. (A sweet and crunchy one, like you would eat—not like the cookie that you have to accept when visiting a web page.)

When I think of a cookie, I think of the jar my mother kept on our yellow Formica kitchen counter throughout my childhood. It was a large porcelain jar with a wide mouth, and often it was filled with homemade cookies. The porcelain lid clanked loudly every time a kid opened the jar for a snack. If I heard my little brother opening the jar, I wandered into the kitchen to get a cookie too. My brother did the same if he heard me. It was a mutually beneficial system—until we got to the last cookie.

When there was only one cookie left in the jar, my brother and I bickered about who got it. It was inevitable. My brother and I squabbled about everything as kids. (As adults, we work in adjacent fields, and there's still a fair bit of good-natured back and forth.) At the time, our cookie conflicts seemed high-stakes. There were often tears. Today, as a parent myself, I admire my mother for stepping in hundreds of times to resolve these kinds of kid disputes. I admire her more for the times she didn't step in and let us work out the problem on our own.

If this story were a word problem in an elementary school math workbook, the answer would be obvious. Each kid would get half of the cookie, or 50 percent. End of story. This mathematical solution is how a computer would solve the dispute as well. Computers are machines that do math. Everything computers do is quite literally a computation. Mathematically, giving each kid 50 percent is fair.

In the real world, when kids divide a cookie in half, there is normally a big half and a little half. Anyone who has had a kid or been a kid can tell you what happens next: there's a negotiation over who gets which half. Often, there is more arguing at this point, and tears. If I was in the mood for peaceful resolution when I was a kid, I would strike a deal with my little brother. "If you let me have the big half, I'll let you pick the TV show that we watch after dinner," I'd offer. He would think for a moment and decide that sounded fair. We would both walk away content. That's an example of a socially fair decision. My brother and I each got something we wanted, even though the division was not mathematically equal.

Social fairness and mathematical fairness are different.

Computers can only calculate mathematical fairness.

This difference explains why we have so many problems when we try to use computers to judge and mediate social decisions. Mathematical truth and social truth are fundamentally different systems of logic. Ultimately, it's impossible to use a computer to solve every social problem. So, why do so many people believe that using more technology, more computing power, will lead us to a better world?

The reason is technochauvinism. Technochauvinism is a kind of bias that considers computational solutions to be superior to all other solutions. Embedded in this bias is an a priori assumption that computers are better than humans—which is actually a claim that *the people who make and program computers are better than other humans*.

Technochauvinism is what led to the thousands of abandoned apps and defunct websites and failed platforms that litter our collective digital history. Technochauvinist optimism led to companies spending millions of dollars on technology and platforms that marketers promised would "revolutionize" and digitize everything from rug-buying to interstellar travel. Rarely have those promises come to pass, and often the digital reality is not much better than the original. In many cases, it is worse. Behind technochauvinism are very human factors like self-delusion, racism, bias, privilege, and greed. Many of the people who try to convince you that computers are superior are people trying to sell you a computer or a software package. Others are people who feel like they will gain some kind of status from getting you to adopt technology; people who are simply enthusiastic; or people who are in charge of IT or enterprise technology.[1] Technochauvinism is usually accompanied by equally bogus notions like "algorithms are unbiased" or "computers make neutral decisions because their decisions are based on math." Computers are excellent at doing math, yes, but time and time again, we've seen algorithmic systems fail at making social decisions. Algorithms can't sufficiently monitor or detect hate speech, can't replace social workers in public assistance programs, can't predict crime, can't determine which job applicants are more suited than others, can't do effective facial recognition, can't grade essays or replace teachers— and yet technochauvinists keep selling us snake oil and pretending that technology is the solution to every social problem.

The next time you run into a person who insists unnecessarily on using technology to solve a complex social problem, please tell them about cookie division and the difference between social and mathematical fairness. You may also want to throw in some comments about how equality is the not the same as equity or justice, how the power dynamics between me and my much-younger brother are a microcosm of power dynamics more broadly, how power in tech rests with those who write and own the code, and how privilege functions vis-à-vis differential access to technology (and cookies).

This book looks at the many ways fairness and bias manifest inside technological and social systems. It offers a starting point for sorting through the many ways that ideas about race, gender, and ability are embedded in today's technology. Digital technology is wonderful and world-changing; it is also racist, sexist, and ableist. For many years, we have focused on the positives about technology, pretending that the problems are only glitches. Calling something a glitch means it's a temporary blip, something unexpected but inconsequential. A glitch can be fixed. The biases embedded in technology are more than mere glitches; they're baked in from the beginning. They are structural biases, and they can't be addressed with a quick code update. It's time to address this issue head-on, unflinchingly, taking advantage of everything we know about culture and how the biases of the real world take shape inside our computational systems.[2] Only then can we begin the slow, painstaking process of accountability and remediation.

In the recent past, critical internet studies scholars Safiya Noble and Ruha Benjamin have used glitches to illuminate the ways that race and technology intersect in pernicious ways. Noble, commenting on the case where Google tagged photos of Black people as gorillas, writes that "algorithmic oppression is not just a glitch in the system but, rather, is fundamental to the operating system of the web."[3] Benjamin builds on this with a concept of the New Jim Code, in which new technologies reproduce and exacerbate historical inequality while being portrayed as neutral or progressive. She writes: "This is more than a glitch. It is a form of exclusion and subordination built into the ways in which priorities are established and solutions defined in the tech industry."[4] Starting from this foundation, I look at a range of incidents reported by journalists (myself included) that show how long-standing social problems are reproduced and amplified inside algorithmic systems.

Using stories of heroes, activists, and ordinary people who have bravely called out bias in tech, I offer a range of straightforward solutions that could make technology better. Sometimes we can address the problem by making the tech less discriminatory, using a variety of computational methods and checks. Sometimes that is not possible, and we need to not use tech at all. Sometimes the solution is somewhere in between.

Coping with the uncomfortable fact that technology discriminates is remarkably like the coping skills required to live in a racist, sexist, ableist society. The book comes from my work as a writer and a computer scientist, as well as my involvement in the wide-ranging field known as artificial intelligence (AI) ethics. I'm a professor of data journalism at New York University. Data journalism is the practice of finding stories in numbers and using numbers to tell stories. I talk, I write, I build software, I run data experiments, I explain carefully what I am doing, and I teach people how to do the things I'm doing. My disciplinary home is in journalism, but I work in an interdisciplinary manner. I do a lot of public speaking about tech and social issues, often under the umbrella of what's called data ethics, critical technology studies, or public interest technology. My goal is always to help people expand their thinking about the future of tech and society.

One of the examples I rely on to explain how tech is biased is the case of the racist soap dispenser. It is a good example of why tech is not neutral, and why the intersection of race and technology can reveal hidden truths. The racist soap dispenser first bubbled up into public consciousness in a 2017 viral video. In it, a dark-skinned man and a light-skinned man try to use an automatic soap dispenser in a men's bathroom. The light-skinned man goes first: he waves his hand under the soap dispenser and soap comes out. The dark-skinned man, Chukwuemeka Afigbo, goes next: he waves his hand under the soap dispenser and . . . nothing happens. The viewer might think: It could have been a fluke, right? Maybe the soap dispenser broke or ran out of power, exactly at that moment? Afigbo gets a white paper towel, shows it to the camera, and waves it under the soap dispenser. Soap comes out! Then he waves his own hand under the sensor again, and again nothing comes out. The soap dispenser only "sees" light colors, not dark. The soap dispenser is racist.

To a viewer with light skin, this video is shocking. To a viewer with dark skin, this video is confirmation of the tech bias they have struggled with for years. Every kind of sensor technology, from facial recognition to automatic faucets, tends to work better on light skin than on dark skin. The problem is far more than just a glitch in a single soap dispenser. This problem has its historical roots in film technology, the old-school technology that computer vision is built on. Up until the 1970s, dark skin looked muddy on film because Kodak, the dominant manufacturer of film-developing machines and chemicals, used pictures called "Shirley cards" to tune the film-processing machines in photo labs. The Shirley cards featured a light-skinned white woman surrounded with bright primary colors. Kodak didn't tune the photo lab equipment for people with darker skin, because its institutional racism ran so deep. The company began including a wider range of skin tones on Shirley cards in the 1970s. While this was the decade in which Black stars like Sidney Poitier rose to greater prominence, the change wasn't the result of activism or a corporate diversity push. Kodak made the change in response to its customers in the furniture industry. The furniture manufacturers complained that their walnut and mahogany furniture looked muddy in catalog photographs. They didn't want to print color catalogs, switching from their previous black and white catalogs, unless the brown tones looked better. Kodak's sense of corporate responsibility manifested only once it stood to lose money from its corporate clients.

Most people don't know the history of race in technology, and sometimes they blame themselves when technology doesn't work as expected. I think the blame lies elsewhere. I would argue that we need to look deeper to understand how white supremacy and other dangerous ideas about race, gender, and ability are embedded in today's technology. Once we acknowledge this, we can reorient our production systems in order to design technology that truly gets us moving in the direction of a better world.

This process starts with recognizing the role that unconscious bias plays in the technological world. I don't believe that the people who made soap dispensers *intentionally* set out to make their soap dispensers racist. Nor do I think that most people who make technology or software get up in the morning and say, "I think I'll build something to oppress people today." I think that what happens is that technology is often built by a small, homogenous group of people. In the case of the soap dispensers, the developers likely were a group of people with light skin who tested it on themselves, found that it worked, and assumed that it would therefore work for everyone else. They probably thought, like many engineers, that because they were using sensors and math and electricity, they were making something "neutral." They were wrong.

This book is about the many ways that technochauvinism has an adverse effect on the creation of technology, and it offers ways to intervene. Unfortunately, there isn't a magic switch that we can flip to make technology work for everyone, everywhere. Eliminating racism and ableism and gender bias from technological systems is a complex problem. Race as a single subject is complicated. So is technology, and so are gender and disability. The intersection of all of these concepts is hard to understand. So much so that it needs the space of a book to untangle the threads.

I write this book as a Black woman, which matters because of standpoint epistemology, which is the idea that one's own standpoint will color how you interpret the world. Black people have a different take on power than the white people who have dominated the tech narrative to date. As Howard University professor, Pulitzer Prize winner, and MacArthur Fellow Nikole Hannah-Jones once observed: "Black Americans are amongst the most astute political and social observers of American power because our survival has and still depends on it."[5] I've been thinking about these issues since I was an undergrad studying computer science at Harvard, where I was one of only six women majoring in the field. When I was a professional computer scientist, nobody in upper management looked like me. I've thought about it almost every day since, when I look around to see I'm still one of the only Black women in AI, in data journalism, or at tech conferences. My previous book, *Artificial Unintelligence*, was one of the first and only trade books about artificial intelligence written by a Black woman.[6] It was published in 2018, more than sixty years after the field of AI was created at a meeting held at the Dartmouth math department in 1956. Why don't we have more books by and about AI written by Black women? The short answer is racism, sexism, and structural inequality. The longer answer requires you to keep reading. You'll bring your own intersectional identity to this discussion, and I'll hope to hear from you about what you think from your own perspective.

Our journey begins with some definitions, so we can be on the same page when talking about specific aspects of AI and other modern computational tools. Then we'll look at facial recognition, a technology that should be banned in policing. We'll examine the justice system in general and how it has failed to use technology well, and why using machine fairness in law enforcement will merely reinforce systems of white supremacy. Next is a section about learning: why we shouldn't use computers to assign imaginary grades to students, and how we can educate ourselves to become more inclusive of people with disabilities. We'll talk about gender, which is socially constructed, and why the next frontier for gender rights is inside databases. After gender, we'll turn to tech and medicine. We'll look at why some medical diagnostic systems are racist, and explore why these systems need to be fixed before we start building any AI diagnostic systems. To test today's AI-based diagnostic systems, I ran my own mammograms through an open-source cancer detection AI to find out if it thought I had breast cancer. (Spoiler: the human doctors saved my life. The AI didn't.) This kind of creative system testing is one of the things that I do as an occasional software auditor, a discipline that I argue shows some promise for intervening in the many fairness problems embedded in today's technology. The final chapters build on the success of algorithmic auditing to focus on additional suggestions for positive, forward-looking developments in policy, in software development, and in everyday life.

Many computer scientists have come around to the idea of making tech "more ethical" or "fairer," which is progress. Unfortunately, it doesn't go far enough. We need to audit all of our technology to find out *how* it is racist, gender-biased, or ableist. Auditing doesn't have to be complicated. Download a screen reader and point it at your Twitter feed, and it's easy to understand why social media might exclude a Blind person from participating fully in online conversations. If a city moves its public alert system to social media, the city is cutting off access to a whole group of citizens—not just those who are Blind, but also those who lack technology access because of economics or a host of other reasons. We should not cede control of essential civic functions to these tech systems, nor should we claim they are "better" or "more innovative" until and unless those technical systems work for every person regardless of skin color, class, age, gender, and ability.

Because the technological world is the same as the social world, policies in the tech world need to change too. By "policies," I mean the formal and informal rules of how technology works. Tech companies need better internal policies around content moderation. Local, state, and federal laws need to be updated to keep us all safer in the technical realm.

So, how do we get there? We have options. In the case of the racist soap dispenser, the engineering team would have caught the problem during the most minimal testing if they had had engineers with a wide variety of skin tones on their team. The available evidence suggests that they did not. Google's 2019 annual diversity report shows that the global tech giant has only 3 percent Black employees. Only 2 percent of its new hires that year were Black women. Black, Latinx, and Native American employees leave Google at the highest rates, suggesting that the company's climate and its pipeline need serious work. This isn't a secret: Google made headlines in 2020 for firing Timnit Gebru, the world's most prominent AI ethics researcher and one of its few Black female employees. Google also fired Meg Mitchell, Gebru's ethical AI co-lead. There are other high-profile Silicon Valley whistleblower cases like Ellen Pao or Susan Fowler that are proxies for hundreds of similar but not publicized incidents of sexism, racism, and the intersection thereof.

None of the major tech companies is doing better. Google is typical of the "big nine" tech companies, and in fact Google's inclusion, diversity, and technology ethics efforts are among the industry's best. However, the problem is that their "best" is still pretty bad. Facebook launched a high-profile effort to hire more women that in 2021 actually resulted in a slight decline in the number of female workers at the company.[7] Building nonsexist, antiracist technology requires a different mindset and requires wrestling with complex social issues while also solving thorny technical challenges. The good news: everyone who reads this book will be better able to anticipate how race, gender, ability, and technology intersect to potentially cause oppression. Instead of technochauvinism, I'm going to offer a different solution: using the right tool for the task. Sometimes the right tool is a computer. Sometimes it's not. One is not better than the other. Just like the 50 percent split of a cookie is unrealistic, and the bigger half of the cookie is not necessarily better than the smaller half of the cookie. The better choice depends on your motivation and your needs.

2

Understanding Machine Bias

Before we start talking in more detail about the intersection of technology and social issues, I want to pause and level the playing field by offering some definitions. If you are a person who is an expert in technology, you may have questions about how social bias operates. If you are an expert in social science, you may have questions about what is happening behind the scenes in today's technology. You may have questions about everything social and everything technical, which is perfectly appropriate. We'll start by looking at AI because it is both the most popular and the most widely misunderstood term in tech—a field that is just brimming with complicated and poorly articulated terms. If you read *Artificial Unintelligence*, consider this a refresher course and skim to the next chapter. If you didn't, buckle in.

Let's start by talking about what's real and what's imaginary about artificial intelligence. Most of what you see in Hollywood is fictitious. *The Terminator, Her, Ex Machina, Star Trek, Star Wars*— these are all wonderful, creative visions of possible worlds, but they are all imaginary. Science fiction is not a blueprint for the future. Fictional other worlds and utopian/dystopian futures have been a source of inspiration for many people (myself included), but it is important to stay within the realm of the real when we're talking about social problems and tech futures. The imaginary kind of AI is known as general AI. That's the AI that will take over the world, the so-called singularity where robots become uncontrollable and irreversible— killer robots—and so on. It's not real. Real AI, which we have and use every day, is known as narrow AI. It's math. Lots of people like to imagine that narrow AI is a path to general AI, but it is not.

Another way of thinking about it is that artificial intelligence is a subfield of computer science, the same way that algebra is a subfield of mathematics. Inside artificial intelligence, there are additional subfields like machine learning, expert systems, natural language generation, or natural language processing. Machine learning is the subfield of AI that is most popular right now, along with its own subfields deep learning and neural nets. These are poorly chosen names because it sounds like there's a brain inside the computer. There isn't! Machine learning is computational statistics. Neural nets are named after the neural processes that happen inside a brain, but they do not actually mimic brain function. In deep learning, the computer doesn't actually "learn" anything the way an organic being would. It merely detects patterns in data.

Most of the action with AI in the world right now centers on machine learning, so let's look briefly at how machine learning works. When you use machine learning, you take a bunch of historical data and instruct a computer to make a model. The model is a mathematical construct that allows us to predict patterns in the data based on what already exists. Because the model describes the mathematical patterns in the data, patterns that humans can't easily see, you can use that model to predict or recommend something similar. This is machine learning. The "learning" part happens when the machine derives the mathematical patterns. The more data you put into the model, the more sophisticated the model becomes, and usually its predictions or recommendations become more precise. We say that the model is trained on the data. There are three methods of training: supervised learning, unsupervised learning, and reinforcement learning. In supervised learning, you tell the computer what features of the data are most important for making a prediction. In unsupervised learning, you let the computer pick which features of the data are most important. In reinforcement learning, you let the computer make a choice and tell the computer whether the choice is good or bad, and the computer eventually learns which types of choices the instructor prefers.

A machine learning model is often described as a black box because the math inside it is extremely complicated and is hard to talk about. Calling it a black box allows us to abstract out what is happening and not get bogged down in complex mathematical conversations at work or in our social lives. However, it's important to remember that just because we use the term "black box," it's not that it's *impossible* to describe what happens inside the model. The ability to explain it is limited by the descriptive abilities of the people in the conversation, the context of the conversation, the mathematical background of the people in the conversation, and the imaginative ability of the people in the conversation. This stuff is hard to understand, and it takes focus and a little bit of work. It's well within the reach of everyone, but I won't lie and pretend it doesn't take effort. Let's dig in.

Math allows us to describe relationships between different things using what mathematicians call variables. One of my favorite examples of describing relationships comes from Solon Barocas, a scholar at Microsoft Research, who uses the relationship between credit score and length of employment to demonstrate how a financial firm might want to use machine learning to predict whether to give someone a loan. We'll use Barocas's excellent graphics to walk through a hypothetical scenario.

Let's imagine that you want a loan. When you walk in to an unfamiliar bank or apply online for a loan—whether it's a mortgage, a student loan, or a home equity line of credit—the bank doesn't know you. The loan officers will be uncertain about how likely you are to pay back your loan, so they have to make an educated guess about the likelihood that they will lose the money they are loaning to you. This is called calculating risk. There are many factors that could be used to calculate risk relative to any one person. Let's say the bank has decided to look at just two variables, credit score and length of employment, to decide whether you will get a loan. Different kinds of relationships between the two variables are going to determine who gets a loan and who doesn't.

In our first example, the data scientists at the bank have decided that there is a positive correlation between length of employment and credit score. A correlation is a relationship between two variables that is judged to be more than just chance. The relationship between the variables is said to be statistically significant. Let's dredge up our middle-school mathematics and express this on a graph as a line. If your credit score goes up with the length of your employment, we would call this a linear relationship. If we put length of employment on the x-axis, and credit score on the y-axis, and we use a line that starts at the bottom left and trends upward in a straight line, it will show a linear relationship between the two variables (figure 2.1).

However, a linear relationship is not the only kind of possible relationship. We might use a computer to look at the credit scores and length of employment of 10,000 people and put the points on a graph and discover that a straight line does not sufficiently describe the relationship between how long someone works and what their credit score is. The relationship might be a curve instead of a line. A nonlinear relationship might look like a curve that starts at the bottom left of our same x-y graph and curves upward to the right, as shown in figure 2.2. A nonlinear pattern might mean that young people who have just started their careers have low credit scores because they don't have any credit history, whereas people who have worked steadily for a long time and have paid their bills on time for many years have higher credit scores. We might say that if you fall above the curve you will get your loan, and if you fall

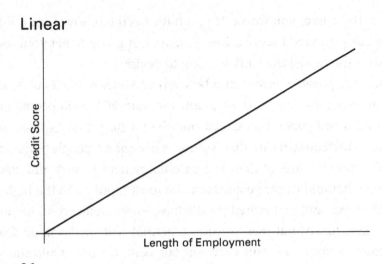

Figure 2.1
Linear relationship between variables.
Source: Solon Barocas and Chandler May.

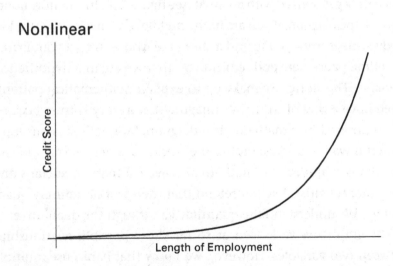

Figure 2.2
Nonlinear relationship between variables.
Source: Solon Barocas and Chandler May.

below the curve, you won't. If you have been employed for a long time but your credit score is low, you are not going to get your loan if this is the model the bank is using to decide.

Another possible relationship between variables is a bell curve, aka a nonmonotonic shape. If we graph the data of 10,000 people and discover a bell curve, we can tell ourselves a story that explains the shape. This could mean that young, early-career people have low credit scores because of their lack of employment history and credit history, that mid-career people have the most money and the highest credit scores, and that retired people have lower credit scores because they have maxed out their earning potential and are living on fixed incomes. A nonmonotonic curve on our same x-y axis would start at the bottom left, swoop up in a curve to the midpoint of the x-axis, then swoop down in a curve that mirrors the upward curve. A bell curve looks like the lump that arises on a cartoon character's head when they get bonked with an anvil (see figure 2.3). In this nonmonotonic shape diagram, if you are in the middle of your career with a low credit score, sorry—you're under the curve, you won't get your loan.

Let me pause here and remind you that we are in a hypothetical scenario. The stories we make up to explain mathematical patterns often have a ring of truth. We imagine that a story is truer because it is supported by a mathematical diagram. Storytelling is our most powerful way of understanding the world on a deep level, and stories play a major role in mathematics even if mathematicians and computer scientists like to pretend that their work is entirely quantitative. We understand the quantitative through the qualitative.

The graphs we've looked at so far show possible relationships between two variables. However, we know that banks use multiple variables when they are deciding if you should get a loan. What does it look like when we add another variable into the mix? This is called a multidimensional relationship, and it can be characterized using the same mathematical terms as above. Figure 2.4 shows a multidimensional linear relationship between three variables:

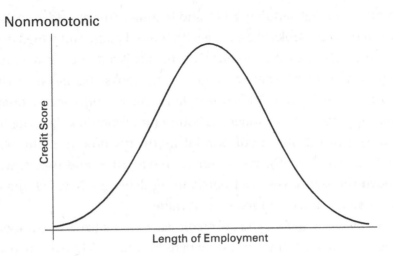

Figure 2.3
Nonmonotonic relationship between variables.
Source: Solon Barocas and Chandler May.

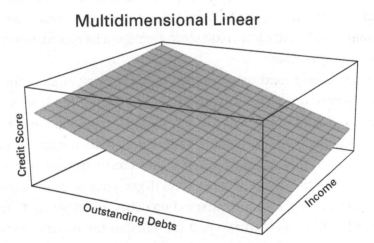

Figure 2.4
Multidimensional linear relationship between variables.
Source: Solon Barocas and Chandler May.

credit score, outstanding debts, and income. This multidimensional linear diagram looks like a cube with a tilted plane embedded in the middle. Like the other graphs shown, this is a model. It is a model representation of what the relationship looks like between three variables taken from 10,000 people in our hypothetical dataset of loan applicants. The model has gotten more complex than the ones we recognize from middle-school math. It's now into the realm of high-school math. We no longer have just x- and y-axes, we've added the z-axis. Let's call outstanding debts the X value, income the Y value, and credit score the Z value.

To predict whether you are credit-worthy, the bank might look at your X, Y, and Z values and see where they fall relative to the plane. People whose scores fall at or above the plane might be considered credit-worthy, and people whose scores fall below the plane might be considered not credit-worthy. Someone with high income, high credit score, and high outstanding debts might be considered very credit-worthy in this model. Someone with high income, low outstanding debts, and a low credit score would not be considered credit-worthy in this model.

If you have a hard time visualizing the relationships in the diagram, you're not alone. If you can, try following the lines with a finger to figure out where a combination of variables would fall relative to the plane. Geometry, as the mathematician Jordan Ellenberg says, is something we often understand with our bodies.[1] If you wanted to predict whether you'll get your loan as someone with low income, medium outstanding debts, and high credit score, follow the lines on each axis and see whether the point is above or below the plane. This is the same procedure that the bank might go through to evaluate you based on your scores—except that the bank will use a computer, not a finger.

Let's get fancier with our model. What does it look like if our three variables of credit score, outstanding debts, and income have a nonlinear relationship? We're no longer looking at a flat plane

Multidimensional Nonlinear

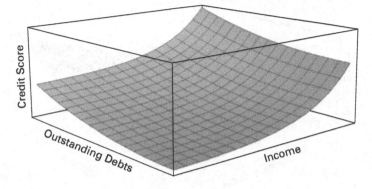

Figure 2.5
Multidimensional nonlinear relationship between variables.
Source: Solon Barocas and Chandler May.

intersecting a cube, but a curved plane inside a cube, as shown in figure 2.5. Again, try to use a finger to figure out what the relationship is. If your three scores fall above the curved plane, you get your loan. If your scores fall below, you're out of luck.

The math is starting to get complicated. If it feels like a challenge, hang in there for just a few more paragraphs. Let's get even more fancy and look at an image that shows a nonmonotonic relationship between our three variables (figure 2.6). The shape inside the cube looks like a shade sail, a kind of canopy you might hang in the backyard to protect yourself from the sun. In this case, though, you don't want to be under the sail because the people who fall under the sail don't get a loan. The space under the sail is more irregular, showing that there is more complexity in the relationship between the three variables.

Figure 2.6 is a relatively simple diagram. Monotonicity doesn't have to be simple, however. A multidimensional nonmonotonic relationship might look like a piece of crumpled paper stuffed inside the cube, as shown in figure 2.7. We've just run up against

Multidimensional Nonmonotonic

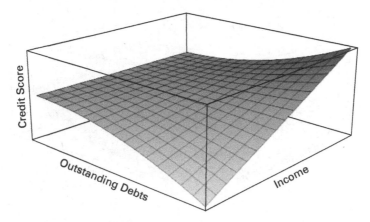

Figure 2.6
Multidimensional nonmonotonic relationship between variables.
Source: Solon Barocas and Chandler May.

Multidimensional Nonmonotonic

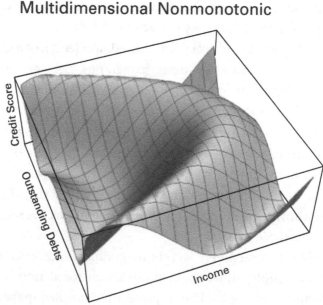

Figure 2.7
Multidimensional nonmonotonic relationship between variables.
Source: Solon Barocas and Chandler May.

the limits of the average human brain. We've plopped into the realm where it's difficult to follow along with a finger. That's okay! Visualizing and manipulating objects in space is an imaginative task that some people can do effortlessly, and some can't—much like rolling one's tongue.

The ability to visualize is also limited by physical constraints. If you are Blind or have low vision, for example, it might be harder to understand the shapes I've shown above. If you are listening to this text instead of reading, you likewise won't see them on the page and will have to create them in your mind.

The human brain is limited in its capacity to conceptualize certain things, and everyone's brain operates a little differently. Imagining big numbers is beyond the capacity of almost everyone. Most people can't imagine what a million or a billion of any object looks like.[2] After a certain point, we all get innumerate. This is why we have machines that do math for us. A computer can easily evaluate whether a person's scores are above or below the crumpled-paper plane in figure 2.7. Even more incredibly, a computer can construct a model that includes more than three dimensions, which is a space that is so complicated that I'm not going to bother trying to describe it.[3]

Do you happen to have a dataset that includes millions of people and contains hundreds or thousands of data points for each person? A computer will slurp up the data, make a model, and make predictions. Machine learning can feel like a neat trick because it takes data from the past and uses it to predict the future. Deciding yes or no on a loan is a kind of prediction. The bank is predicting whether you'll pay back the money they loan you, based on what you and/or people like you have done in the past. The statisticians at the bank look at variables, decide where there is a statistically significant correlation between variables, and use the resulting math to determine if you get a loan. This basic process is also how Spotify predicts what you'll want to listen to, how Google predicts what you're searching for, how Pinterest and Instagram predict what images you'll like looking at, or how Twitter and Facebook predict which posts you'll want to read or engage with. For visual data, the process works in a similar way. If you have a bunch of photos, you ask a human to identify the object by drawing a bounding box around it and applying a text label. Amazon offers a service called Mechanical Turk that allows data scientists to pay people tiny amounts of money to draw bounding boxes and apply labels to hundreds of millions of photos and videos. Those human-labeled examples are then fed into the computer. The computer will construct a model that tracks the arrangement of pixels and predicts a bounding box and label for collections of pixels that are mathematically similar to the labeled areas. This is how a computer can identify a face, a cat, or a motorcycle in a photo. The computer uses math on the pixels, which are in a grid on an x-y plane. A human has already set up the architecture identifying a collection of pixels as a cat or whatever.

Whoever owns the model has an enormous amount of power. Labeling a group of pixels as "cat" is not a given. You could easily label it as "crumpled paper" and nobody would likely notice. This is one of the potential dangers of large models. They contain millions of data points and even more associations, but nobody is checking on each one of them, so errors and biases can flood the system.

For generating language, the process is also similar. GPT-3, the most popular language-generating model right now, is trained on millions of books and millions of social media posts. It slurped up the data (from Reddit, Google Books, and other sources) and found the patterns in words and in which words follow each other. It's better than its precursor, GPT-2, because it is trained on more data.

Data, model, prediction, math. That's the core of machine learning. It's not magic. It is an impressive human achievement, and the math underneath it is beautiful, but it is not magic. It also does not have any larger transcendental meaning. We are not entering into a new phase of human evolution because we can do more math with machines. Math will not save us from mundane things like having to do the dishes. Math can help us make machines that wash the dishes, but the dishes will still need to be done and the machine will inevitably break, need repairs, or need replacing. We can't let the fantasy distract us from the daily realities of using machines.

These machines, the computers, are produced out of a specific human context. So is the data we use to train the machine learning models. This causes problems. We're now in the realm of what's called AI ethics or responsible AI or data ethics. Critical race and digital studies (CRDS) is a field that looks at the ways race, ethnicity, and identity shape and are shaped by digital technology. CRDS work by scholars such as Alondra Nelson, Ruha Benjamin, Safiya Umoja Noble, Charlton McIlwain, Catherine Knight Steele, Lisa Nakamura, André Brock, Sarah J. Jackson, or Meredith Clark provides a bridge for understanding the intersection of technology and race, ethnicity, or identity.

Many people, when confronted with ableism, race, or gender bias in tech, tend to consider it a glitch. A glitch is something temporary, a mysterious blip that may or may not be repeated. A bug is a more serious matter that makes the software fail, and it is worth addressing. A glitch is ephemeral and can be dismissed; meanwhile, a bug is substantial, ongoing, and deserves attention. Developers use cognitive shortcuts to figure out which problems merit fixing. This is a normal strategy—it's how the human brain operates. We all use cognitive shortcuts, and in today's world where decision making is more complex than ever, shortcuts are essential. The problem is, shortcuts often contain bias. Shortcuts focus on race, gender, ability, and other superficial categories. Sometimes those shortcuts are innocuous or helpful. A doctor uses cognitive shortcuts to diagnose patients. If a woman in her forties presents with persistent exhaustion, the doctor will check her thyroid levels because it's quite likely to be a thyroid issue. It doesn't make sense to check for, say, broken bones. Going to the most likely explanation saves time and is in this case helpful. On the other hand, cognitive shortcuts are often based on problematic assumptions. A white woman who clutches her handbag close when a Black man passes by is using a cognitive shortcut that considers the Black man a likely handbag-snatching thief. This is a racist assumption. And, as we know, small bits of racism spiral out into larger policies of racism that exacerbate inequality and lead to violence against entire groups of people.

Cognitive shortcuts are tested by edge cases, things that don't fit neatly into existing mental models. In computing, an edge case is something that is outside the normal realm of what is addressable by a computer. The idea is, you design a program for the greatest number of situations, but there will always be edge cases that are weird or fall outside the realm of normal. The problem is, "normal" is usually whatever the developer's own experience is. People also consistently overestimate how much of the world is made up of people like themselves. This is why so much technology is optimized for able-bodied, white, cis-gender, American men, because this population makes up the majority of software developers. Whiteness is perceived as the invisible default for software system users, while any racial or ethnic affiliation other than white is perceived as visible or aberrant. Visibility translates to being outside the realm of normal—an edge case.

Humans are often good at dealing with edge cases, but computers are not. The law is designed to accommodate everyone, yet when the law is embedded in algorithms it is often used to prosecute, exclude, or marginalize groups that particular tech developers have labeled edge cases. People experience real harms committed by technological systems. Algorithmic accountability is the field that evaluates or audits algorithmic systems and holds firms accountable for the results. Algorithmic accountability is sometimes considered journalism, sometimes considered AI ethics, sometimes considered public interest technology. The fact that algorithmic accountability exists as a field and a practice is important. The label, less so. The Markup, ProPublica, the *Wall Street Journal*, and the *New York Times* are some of the news organizations best known for in-depth algorithmic accountability reporting. These kinds of report-ing projects are extremely expensive and time-consuming and require teams of people to achieve.[4]

Because the math used in AI for determining loan eligibility is so complex, and the numbers are so large, it can make it hard to under-stand (much less contest) the structural biases written into code. However, the results of ignoring the problem cause real harm: Black applicants are turned away more frequently than white applicants, Black applicants are offered mortgages at higher rates than white counterparts with the same data, and there are examples of every other possible bias and consequence.[5] Substantive reform is needed. A number of mathematicians have begun pushing for math to be used for good, not evil. Mathematicians have gotten involved in the gerrymandering problem, in which legislative districts have been strategically divided to benefit one party. There was a call for math-ematicians to boycott collaborating with police in the wake of the murders of George Floyd, Breonna Taylor, Tony McDade, and oth-ers. "Given the structural racism and brutality in US policing, we do not believe that mathematicians should be collaborating with police departments in this manner," stated a boycott letter with 1,930 signa-tories in the June 2020 Notices of the American Mathematical Soci-ety. "It is simply too easy to create a 'scientific' veneer for racism."[6]

We should talk briefly about race and racism and antiracism. Race is socially constructed. When I say "racism," I mean a kind of bias based on perceptions of racial groups. Race is different than ethnicity, and both genetics and epigenetics are different than race. The so-called racial groups that persist today were manufactured in the 1400s in order to justify slavery. If some groups were greater and some lesser, enslavers could justify their despicable trafficking in human bodies. Enlightenment philosophers like Carl Linnaeus dug in on this idea. Linnaeus created classification schemes that separated humans into hierarchical groups based on their continents of origin, again claiming that some origins were superior to others. He also is credited with developing the idea of taxonomy, which we still use today to categorize the natural world. Linnaeus gave us the term *Homo sapiens* and codified scientific racism. His ideas on taxonomy were so helpful in classifying plants and animals that his racist ideas about groups of people were also accepted as fact. In this we can see the origins of technochauvinism, where computer scientists create such useful tech systems that their problematic ideas about society are overlooked. Race is a social construct but it is often embedded in computational systems *as if it were scientific fact.*

Of course, there are nuances to understanding race inside sociotechnical systems. It is occasionally useful to track racial statistics to ensure equitable access for different groups of people. Keeping track of racial statistics in school admissions allows us to see school segregation. In 2021, only eight Black students were accepted to New York City's Stuyvesant High School, one of the country's most elite public schools. That's clearly a problem in a class of 749. Even more troubling is the fact that less than 4 percent of Black students who applied received offers, while 28 percent of white students did. New York City's school system has a problem with segregation. Racial statistics allow us to articulate the problem.

Will educational equity be achieved through simply showing racial disparities in NYC schools? No. "The master's tools will never dismantle the master's house," Audre Lorde wrote. "They may allow us temporarily to beat him at his own game, but they will never enable us to bring about genuine change."[7] Statistics alone are not sufficient to achieve social change, though they are an important part of an activist's toolbox. Likewise, we cannot depend on computational solutions alone (including social media campaigns) to achieve lasting social change. Computers are important tools in the fight for social justice. People are the ones who power change.

Often, the way racial ideas manifest in technology serves to enable and codify racial discrimination. Algorithmic systems often act in racist ways because they are built using training data that reflects racist actions or policies. In our example above about loans, we would be training our AI model using real-world data about who had been given loans in the past. In the United States, there is a deep history of racial bias in financial services (as well as in housing, policing, and everywhere else). Historically, fewer Black and Brown people have been given loans. If we feed the model data about who has been given loans in the past, the model will continue to reject Black, Indigenous, and people of color (BIPOC) applicants. This is not just theoretical. In a 2021 investigation of mortgage-approval algorithms, The Markup reporters Emmanuel Martinez and Lauren Kirchner showed that loan applicants of color were 40–80 percent more likely to be denied than their white counterparts.[8] Many technochauvinists imagine that data is more objective, that using only data will get us to a fairer and less-biased world. Again, I would argue for nuance. Both data and human systems have problems, and we need to work through all of them.

It is no longer enough to say, "I'm not racist." "In a racist society it is not enough to be non-racist, we must be anti-racist," Angela Y. Davis said. Ibram X. Kendi's bestselling book, *How to Be an Antiracist*, is an excellent starting point for understanding antiracism. Antiracist action and mindset means critically examining one's own beliefs about race and taking corrective action to eliminate racist practices and weed out white supremacy.

Taking Davis's and Kendi's work one step further, we can consider how antiracism can be embedded in our technical choices. Antiracism goes beyond merely implementing the most mathematically fair solution in a machine learning system. It means that while we build technology, we must challenge the systems of oppression that exist in the world, so we don't reproduce that oppression inside our technical systems. Privileging the quantitative over the qualitative, as in technochauvinism, is often part of a system of oppression. Someone who builds a computational system without the intent to be racist or sexist or ableist may still build a system that is biased, perhaps because of the builder's unconscious bias or because the system is built on top of a problematic social structure. This happens often in medicine, as we'll see in a later chapter.

I know that some people like to debate whether racism is real, and whether race is real. Some people like to argue about the gender binary. Some people argue that people with disabilities make up such a small minority that it's not worth factoring their experiences into system designs. This book does not engage with any of these arguments.

Instead of imagining that we can solve every social problem with math, I suggest that we think more holistically. Let's not default to tech when it's not necessary. Let's not trust in it when such trust is unfounded and unexamined. Let's stop ignoring discrimination inside technical systems. Let's talk more about why tech is biased, in order to make the necessary changes that will make tech less biased. If training data is produced out of a system of inequality, don't use it to build models that make important social decisions unless you ensure the model doesn't perpetuate inequality.

3

Recognizing Bias in Facial Recognition

Now that we've discussed the fundamentals of AI and bias, let's look deeper into how race and bias are coded into technology, with disastrous results. On a cold day in January 2020, Robert Julian-Borchak Williams was at work at an automotive supply store when he got a strange call. Detroit police wanted him to go to the precinct and be arrested. Williams, who lives in a suburb north of Detroit, thought it was a prank call. He hadn't committed any crimes; why would someone want to arrest him? Williams told the caller, "Well, I'm leaving work in 15 minutes, so unless you can get here in 15 minutes you can come to my home." He shrugged off the call, finished up his work for the day, and drove home.

Detroit police were waiting at his Farmington Hills house when he pulled into the driveway. A squad car veered up to block Williams's SUV, as if he were going to flee.

"Are you Robert Williams?" an officer asked.

"Yes," said Williams.

"You're under arrest," said the officer. Williams asked why. The officer briefly showed him a paper. Williams saw his own name, with "arrest warrant" and "felony larceny." Williams's wife came outside with the couple's two young daughters. Mrs. Williams held the younger one in her arms and tried to shield the elder daughter behind her legs. Both girls were crying.

"Go back inside," Williams called to his daughters. "Daddy will be back in a minute." He wouldn't. The police handcuffed him, while his wife and two small children looked on, and took him to jail.

Williams sat in the Detroit Detention Center, confused and wondering what was going on. He knew that his civil rights were being violated, and he was livid at a world where being Black was grounds for arrest. "A full 18 hours went by," Williams said later. "I spent the night sleeping on the cold concrete floor of a filthy, overcrowded cell next to an overflowing trash can. No one came to talk to me or explain what I was accused of."[1]

Williams was arrested because of an algorithm. Facial recognition technology (FRT) had wrongly identified him as a suspect in a robbery that happened over a year earlier, in October 2018, when someone stole $3,800 of watches from a Shinola store in midtown Detroit.

The shoplifting episode was business as usual for the police. 28,000 incidents of retail fraud are reported in Michigan every year. After the shoplifting incident, the Shinola store gave a copy of its surveillance video to the Detroit police. Five months later, a digital image examiner for the Michigan State Police looked at the grainy, poorly lit surveillance video on her computer and took a screen shot.[2] She uploaded it to the facial recognition software the police used: a $5.5 million program supplied by DataWorks Plus, a South Carolina firm founded in 2000 that began selling facial recognition software developed by outside vendors in 2005. The system accepted the photo; scanned the image for shapes, indicating eyes, nose, and mouth; and set markers at the edges of each shape. Then, it measured the distance between the markers and stored that information. Next, it checked the measurements against the State Network of Agency Photos (SNAP) database, which includes mug shots, sex offender registry photographs, driver's license photos, and state ID photos. To give an idea of the scale, in 2017, this database had 8 million criminal photos and 32 million DMV photos. Almost every Michigan adult was represented in the database.[3] Each of the photos in SNAP had already been measured for markers. Similar ones were pulled into a pot, and that pot was then checked using machine learning to evaluate mathematical similarity. The similarity score was tallied, and the photos with the highest similarity scores were suggested to the examiner. The program matched the man in the grainy surveillance image, who wore a Cardinals baseball cap, to Robert Williams's driver's license photo. She then sent the photos off to the Detroit police, telling them it was a likely match. The police took the algorithm as truth and arrested Williams.

The Hollywood scenes of algorithmic facial recognition, where the computer flips through millions of faces in an instant and selects the correct one, are more glamorous than the reality. How facial recognition systems really work is a practical, understandable process with many opportunities for mistakes. Facial recognition is known to work better on people with light skin than dark skin, better on men than on women, and it routinely misgenders trans, nonbinary, or gender nonconforming people. Though it tends to work the worst on Black and Brown people, it is often deployed against these communities in policing.[4] In the Williams case, the computer program was wrong—as was the person examining the image, as were the other humans in the police decision chain.

It's also important to note that a facial recognition system does not say it has a definite match. It computes *similarity*, and its chosen match is the most likely match out of all the data it has available. It is only checking against what it has been trained on—in this case, Michigan driver's license photos, state IDs, and so on. As one major issue in this case, the system doesn't even check people from other states. It's odd to assume that people who live in Michigan would be the only ones to commit crimes within the state boundaries, but technochauvinist belief in technology means a lot of blind spots.

Law enforcement does not have a good track record when it comes to using high-tech tools for policing. DataWorks has been in business since 2000, but its successes are few. Kashmir Hill, who broke the Williams story for the *New York Times*, wrote: "In Michigan, the DataWorks software used by the state police incorporates components developed by the Japanese tech giant NEC and by Rank One Computing, based in Colorado. . . . In 2019, algorithms from both companies were included in a federal study of over 100 facial recognition systems that found they were biased, falsely identifying African-American and Asian faces 10 times to 100 times more than Caucasian faces."[5]

There are layers of procedural checks built in: according to policy, the match suggested by the computer has to be validated by a human checker and a supervisor, and the system is not supposed to be used on real-time video feeds.[6] This is supposed to be for safety. Indeed, it's good that there are additional checks built into the system. However, human biases come into play and don't necessarily provide the safety intended. If you get a checker who thinks all Black people look alike, or all Asian people look alike, it's going to be a problem. If you get a checker who is bored with running photos through a computer program and emailing the results afterward, the quality of the checking will decline. There's also the practical concern that the person doing the checking wants to keep their job, and their job is to use the computer program, so they don't have any incentive to say the computer program doesn't work. And there's the insidious possibility of technochauvinism. The human checker may believe that technology holds all the answers and that the system can't possibly be wrong.

The cascading disasters in the Robert Williams case illustrate the many, many problems with facial recognition systems in policing. It's important to understand the way that racial bias is embedded in and perpetuated by these systems, in order to dismantle the architecture of white supremacy that tolerates this and countless other violations of people's civil rights.

Let's start with the alleged incident, the theft of some watches retailing for $3,800. It's not a good idea to prosecute shoplifting that's this low in dollar value. It costs taxpayers upwards of $2,000 each time someone enters the criminal justice system, according to the National Association of Shoplifting Prevention.[7] Unfortunately, retailers' cost-cutting attempts have made it easier to shoplift: Self-checkout stations, for example, have directly contributed to an increase in retail theft. "Initially, self-checkout was thought to be one of those silver bullet kind of solutions to the increasing costs of doing retail, because your biggest expense is usually your labor costs," said Richard Hollinger, a retired University of Florida criminologist who conducts an annual security survey for the National Retail Federation.[8] "But the downside is that it is somewhat easy to steal in from those kinds of locations, particularly if you can scan one item and take two or three items." Stores that invest in automated checkout technology to replace staff are making an unwise economic choice that shifts costs to the state and the taxpayer. Then, the state ends up spending more public money on surveillance and prosecution, because retailers want to incrementally increase profits. This is not a good deal for the public.

Another factor is that Detroit police invested millions in software and were looking for ways to use it. In part, they were looking to justify the taxpayer expense. Project Green Light is an enormous investment in surveillance technology made by the Detroit Police Department. It began in 2016 with the installation of security cameras at eight gas stations. The camera feeds were hooked up to a "real time crime center" that uses facial recognition software and license plate readers, plus has access to GPS data. The FBI and the Department of Homeland Security were involved in the project, as was a retailers' association. "This project is the first public-private-community partnership of its kind, blending a mix of real-time crime-fighting and community policing aimed at improving neighborhood safety, promoting the revitalization and growth of local businesses, and strengthening DPD's efforts to deter, identify, and solve crime," reads the City of Detroit's promotional copy.[9] The police have paid for the software, and so they want to use it—even if it works badly. This mistaken belief is technochauvinism. Such faith in computational results is low stakes when Google Maps or Waze suggests getting off the highway and driving on surface streets when you are trying to get to the airport. But it is high stakes when it's wildly expensive, involves the risk of personal danger in jail, and is going to affect the rest of someone's life. Michigan has a particularly bad track record of failing at enterprise technology—ironic, considering that the state is known for its history of using assembly lines for automotive production, an earlier generation's case study in using technology efficiently.

In the case of facial recognition, technochauvinism intersects with racism in dangerous ways. Looking at the photo, and looking at Robert Williams, the police could have said that the match was too tentative to pursue. They didn't, because every person in the chain of events was predisposed to believe that a Black man would be a thief. When Robert Williams was shown the blurry surveillance photo that had been "matched" to his driver's license photo, he actually laughed a little—the only resemblance was that he and the person in the photo were both Black. The man in the photo wore a Cardinals cap. Williams is not a Cardinals fan; he doesn't even follow baseball.[10] He held the photo next to his face. "I hope you guys don't think that all Black men look alike," he said to the officers. They looked at each other.

"The computer must have gotten it wrong," one grudgingly admitted. Instead of releasing him, however, they kept Williams for several more hours. He was released into a freezing and rainy night, waiting outside in the damp and the cold as his wife arranged childcare for their sleeping daughters so she could drive downtown and pick him up.[11]

In all likelihood, nobody in the police department thought they were being racist. They probably thought they were just doing their jobs, prosecuting crimes and using the department-approved tools. Not being racist is insufficient. Following Ibram X. Kendi's distinction between nonracist and antiracist, it is necessary to be *antiracist* in order to effect change. A nonracist person says, "I'm not racist" and continues on as before, claiming to be neutral on topics of race. An antiracist person, Kendi explained in a 2020 talk, questions the tools and policies involved in policing. "[A] not-racist is a racist who is in denial, and an antiracist is someone who is willing to admit the times in which they are being racist, and who is willing to recognize the inequities and the racial problems of our society, and who is willing to challenge those racial inequities by challenging policy," he said. This is particularly hard because moving from neutrality to antiracism requires being vulnerable. "You have to be willing to admit that you were wrong," Kendi said. "You have to be willing to admit that if you have more, if you're white, for instance, and you have more, it may not be because you are more. You have to admit that, yeah, you've worked hard potentially, in your life, but you've also had certain advantages which provided you with opportunities that other people did not have."[12] The macho, bro culture of Silicon Valley that makes facial recognition software does not value vulnerability. The macho, bro culture of American law enforcement does not value vulnerability. Instead of admitting that the software doesn't work, has never worked, and is never going to work, the software-makers and software-buyers and software-users double down on their mistakes. Innocent, law-abiding citizens like Robert Williams suffer as a result.

As of this writing, Robert Williams is free and has lodged a complaint against the Detroit Police Department with the help of the ACLU. In July 2021, he testified before Congress about his wrongful arrest, as part of a hearing examining the use of facial recognition technology by law enforcement convened by the House of Representatives Committee on the Judiciary Subcommittee on Crime, Terrorism, and Homeland Security.

The reason Williams was released is because of a new public understanding of facial recognition technology brought about by a researcher in Cambridge, Massachusetts, hundreds of miles away. Dr. Joy Buolamwini wanted to make a mirror. It was 2016, she was a graduate student at the MIT Media Lab, and she had a class assignment to make a real-world object inspired by a science fictional future. The Media Lab is a unique place: it is full of state-of-the-art equipment like 3-D printers, as well as eclectic items like pre-release Legos and experimental sneaker soles. Class assignments are sometimes problem sets and sometimes about building things or making art or putting computers into ordinary objects. Buolamwini wanted to make a mirror that would inspire her every morning by delivering positive affirmations. The plan was solid. She would use a new kind of technology that let a thing act as a mirror and also as a computer screen. She named it the Aspire Mirror.

"The development of the Aspire Mirror was influenced by emotional devices like the empathy box and mood organ in Philip K. Dick's science fiction classic *Do Androids Dream of Electric Sheep?* (1968) as well as Anansi the spider, a legendary creature spoken about in Ghanaian tales about shape shifting," Buolamwini said. "Perhaps by seeing ourselves as another, by shifting the shape of our reflections, we could engender more empathy. Staring into the Aspire Mirror, an onlooker could see reflected onto her face animals, quotes, symbols, or anything else we could code into the system."[13] When she looked into it, she wanted the mirror to recognize her face and display a positive affirmation on the screen around her face, overlaid with an image of an inspiring role model like Serena Williams. Her professor was enthusiastic. This was a good use of new mirror technology. The commercial possibilities were exciting too. Who wouldn't want to wake up to a mirror that tells them they are beautiful?

Buolamwini had come to earn a PhD at the Media Lab after earning a graduate degree as a Rhodes Scholar at Oxford. She was also a poet and a former pole vaulter who considered going to the Olympics but decided to focus instead on her education.

Although the project was flawlessly conceived, the problems with the technology started immediately. The inspirational mirror relied on a camera hooked up to a facial recognition program on a computer. Buolamwini put her face in front of the camera and . . . nothing happened. Maybe it was broken? Maybe it was a glitch?

The daughter of a pharmaceutical scientist and an artist, she was not unfamiliar with being the only Black person in the room in tech environments. This was also not the first time that facial recognition had failed to detect her face.[14] Most of her classmates had light skin, and the computer had recognized them. Buolamwini had dark skin. The computer had not recognized her.

White people, confronted with evidence of racism, often don't believe it. "I don't know why it's not working for you" is the typical reaction I've seen in this situation. Buolamwini, who had seen this happen before, took a practical step that changed the course of computational history. She put on a white mask. The computer recognized her. She took off the white mask, and the computer failed to recognize her. She put the mask back on, and it recognized her again. Even the most skeptical people are silenced by this incontrovertible evidence, which Buolamwini demonstrated in a widely-circulated video. The computer didn't recognize Joy Buolamwini because of her dark skin.

Let's talk about the white mask. It was simple: a bright white molded plastic full-face mask. It was the kind of mask favored by the kinds of tech activist characters you might find in the show *Mr. Robot*. The Media Lab is the kind of place where nobody thinks twice about having a white mask in your office. People have all kinds of odd objects in their offices. Buolamwini's advisor had a bed in his office. He slept there a few nights every week because he lived two hours away from Cambridge in Western Massachusetts.[15]

Buolamwini had the mask because of a misunderstanding. A new friend had invited her to a girls' night. "We'll do masks and hang out," the friend said. "Bring a mask." Buolamwini thought it was a strange request, but she's up for new experiences, and she went to the store to buy a mask for what seemed like a costume party. She did not realize the friend meant a facial mask of the kind you might buy at Sephora, that tightens your pores and makes your skin smell like fruit. The misunderstanding was hilarious, and the mask ended up sitting around in her office for months afterward. When she wore the mask, it was easier to code her project. This is the kind of thing that Black people are subjected to when using racist tech. Buolamwini calls it the coded gaze. If she wanted a good grade on her assignment, if she wanted to make the tech work well enough to recognize her, she had to disguise herself with a white mask.

Not long after, Buolamwini went to a book launch event at Harvard Bookstore featuring mathematician Cathy O'Neil, author of *Weapons of Math Destruction*. It changed Buolamwini's life. She got activated, realized there were other people thinking about algorithmic bias, and decided to shift her focus toward algorithmic justice. She collaborated with another AI researcher, Timnit Gebru, now the founder of the Distributed Artificial Intelligence Research Initiative, to write a groundbreaking paper called "Gender Shades" that looked at how several computer programs classify gender. Then, Buolamwini founded the Algorithmic Justice League, a collective dedicated to mediating algorithmic harms.

"Gender Shades" did several important things. First, it raised concerns about facial analysis tasks like gender classification or facial recognition. Second, it showed that insufficient test data prevented evaluators from uncovering bias. Buolamwini and her team didn't stop with this powerful demonstration; they then assembled a new dataset, sourced from international politicians' photos, which has a full range of skin tones for two genders. Using a diverse dataset increased the accuracy of the model results. Many people would consider this a complete solution, just making the training data more diverse and thus increasing facial recognition's accuracy. It is not. The solution, as Buolamwini has said many times, is not in making the model better so that Black and Brown people are more visible to surveillance systems. The socially just decision is to not use facial recognition at all in policing, because these systems are too often deployed to harm communities of color and poor communities. We could make the tech more accurate by using more diverse training data. But it would not actually fix the problem, because it would exacerbate inequality.

"Gender Shades" was such a blockbuster, and so crystal-clear in its conclusions, that the Big Tech firms reeled. Microsoft and IBM halted development of FRT for law enforcement. Amazon called a one-year pause in developing the technology for policing, perhaps hoping that public attention would be diverted in the meantime. (It wasn't.) Amazon announced an "indefinite pause" a year later.

The National Institute of Standards and Technology (NIST) replicated the "Gender Shades" study and validated its findings. A 2019 NIST report noted that commercial facial recognition systems are biased and that they falsely identify Black and Asian faces 10 to 100 times more often than white faces. Native American faces generated the highest rate of errors. Older adults were misidentified ten times more often than middle-aged adults.[16]

Opposition to racist, sexist facial surveillance tech has swelled. As of this writing, seventeen cities have banned facial recognition. Even as opposition grows, the situation gets worse on the commercial side. Half of the federal law enforcement agencies use FRT, according to a July 2021 audit by the US Government Accountability Office.[17] The FRT systems can contain hundreds of millions or billions of photos. Of the seventeen systems purchased by federal agencies, few of them work. In March 2020, only four were in service, and at only three agencies: the Federal Bureau of Investigation, Federal Bureau of Prisons, and US Customs and Border Protection. The patchwork system of state regulations allows for convenient loopholes. If a law enforcement official is prohibited from using FRT themselves, they can ask a colleague who is permitted to do the search for them. Of the forty-two federal agencies surveyed by the GAO, seventeen reported they had used another entity's system with facial recognition technology from April 2018 through March 2020. The FRT systems gather data from a variety of sources: mug shot photos, driver's license photos, passport photos, publicly available photos on the internet, and images from video or closed-circuit television.

The FBI's Facial Analysis, Comparison, and Evaluation Services has memorandums of understanding with several state agencies, which allow it to leverage the state-owned systems for facial recognition searches. This is good that it is documented. Unfortunately,

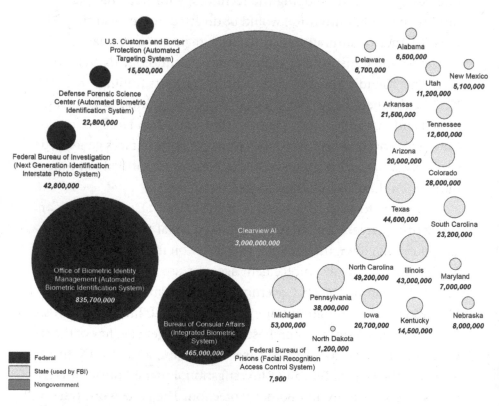

Figure 3.1
Depiction of selected federal, state, and nongovernment systems with facial recognition technology used by federal agencies that employ law enforcement officers, and the number of photos in them, as of March 31, 2020.
Source: GAO-21-105309, https://docs.house.gov/meetings/JU/JU08/20210713/113906/HMTG-117-JU08-Wstate-GoodwinG-20210713.PDF; GAO analysis of information provided by system users or owners.

the actual use of the state-owned systems is unclear. Of the fourteen agencies that reported using nonfederal systems, thirteen do not have a way of tracking what employees are doing with those systems. "When we requested information from one of the agencies about its use of nonfederal systems, agency officials told us they had to poll field division personnel because the information was not maintained by the agency," read one GAO report. "Officials from another agency initially told us that its employees did not use nonfederal systems; however, after conducting a poll, the agency learned that its employees had used a nonfederal system to conduct more than 1,000 facial recognition searches."[18] This could mean using a friend or colleague's account on a commercial facial recognition service like Clearview AI; it could also mean doing a Google reverse image search.

Of the twenty federal agencies that claimed to use FRT, the majority used it for criminal investigations and surveillance. The FBI claimed to use it for investigations of violent crimes, credit card and identity fraud, missing persons, and bank robberies, among others. Six agencies claimed they had used FRT to "support criminal investigations related to civil unrest, riots, or protests." "All six agencies reported that these searches were on images of individuals suspected of violating the law," read the report.[19] though the majority of the protests were legal and peaceful.

FRT played a much-hyped role in identifying the rioters involved in the January 6 Capitol insurrection. It was not the only tool used, but it was effective—because the insurrectionists were foolish enough to post pictures of themselves on social media with their identities attached. Identifying them was like shooting fish in a barrel. It's not high-tech police work. Criminals are rarely witless enough to photograph themselves committing crimes and to brag about it on public websites. Those who were unwise enough to admit to their participation were easily caught. One insurrectionist was turned in to the FBI by a woman who matched with him on Bumble, a dating app. Just after January 6, she searched on the app for conservative men in DC. The rioter replied and said he was involved in the insurrection; she turned him in to the FBI. It was not the only case of citizens using simple consumer tech to find insurrectionist criminals.[20]

As of this writing, facial recognition is still used in policing, despite its poor reputation, and its advocates are offering small concessions to try to placate those who are calling for bans. The *Washington Post* reported: "Defenders of the technology said it should be used solely to generate leads for police, not as the lone piece of evidence, and that officers should not rely too heavily on its results or apply it to every low-level crime. The Detroit department's policy has since been changed to allow the use of facial recognition software only in cases of violent crime."[21] Without a wholesale ban, activists and policymakers are stuck playing whack-a-mole to keep police from using facial recognition. Of the seventeen cities where facial recognition is banned, six of the cities have loopholes that police are exploiting nonetheless.[22]

Robert Williams's story was the first major media episode of someone being wrongly arrested because of an algorithm. Since then, more have emerged, and many more are possible. In May 2019, also in Detroit, a Black man named Michael Oliver was arrested for a crime he didn't commit. In this case, a group of students was fighting. A teacher saw it, called 911, and used his cell phone to record the incident. One of the people in the melee reached into the car and grabbed the phone, throwing it on the ground and cracking the case. The phone video was later fed into a facial recognition program that wrongly identified Michael Oliver as a lead suspect. Oliver was charged with a felony count of larceny. The facial recognition software's choice was particularly absurd because Oliver has tattoos all up and down his arms, and the perpetrator clearly doesn't. Oliver's skin is lighter, and his hair is several inches longer than that of the man in the video. He was arrested and had to pay for a lawyer. The lawyer had the case tossed out because it was a clear misidentification.[23]

Another case is that of Nijeer Parks, also Black, a resident of Paterson, New Jersey, who worked at a grocery store. There was a shoplifting thirty miles away: a man trying to steal snacks from the gift shop at the Hampton Inn in Woodbridge when he was in the lobby extending the rental on a gray Dodge Challenger. The dispute over snack food escalated and two police officers arrived at the scene. The shoplifter's driver's license was found to be fake. He drove off, allegedly trying to hit a police officer in a parked car, then crashing the car into a column in front of the hotel. The man then fled. The police used facial recognition and got a "match" to Parks, who was not involved. As of January 2021, Parks is suing the police, the prosecutor, and the Township of Woodbridge for false imprisonment and violation of his civil rights.[24] These are unlikely to be the last cases. Faulty facial recognition has caused problems in multiple contexts—usually for people with darker skin.

Facial recognition software is far less effective in policing than anyone imagines. Its big successes have been in identifying a couple of shoplifters. I'm not too worried about the public safety menace of shoplifting. Taking a candy bar from a Duane Reade is not the end of the world. We should cut back on police spending on technology that does not actually keep us safer, and we should be extremely critical of claims that technology actually works for policing. Moreover, we should demand accountability and reform for technologies that have not worked since the 1960s and are actually racist. The best way to be antiracist when it comes to facial recognition is not to use the technology at all in policing. The need to enact reform in police technology doesn't stop with facial recognition, however. As we'll see in the next chapter, facial recognition algorithms are not the only problem that we have to worry about in policing.

4

Machine Fairness and the Justice System

Precognition is attractive. "If only I had known" is what we often say after something horrible happens. There's a sense that if we could get advance knowledge, we would be spared pain and regret and tragedy. We would do something different. We would avoid pain for ourselves and our loved ones. Makers of so-called crime prediction technology lure public officials by promising statistical and machine learning methods that will predict where crime will happen and who will commit it, so that police can intervene before the bad things happen. But this technology is biased against Black people, Brown people, under-resourced people, and LGBTQIA+ people. In other words, anyone who is not a wealthy white American cisgender heterosexual male should beware. A closer look at machine fairness reveals that using machines for predictive policing is as effective as pouring salt on a wound.

People have varied skill levels when it comes to using technological tools. The police are no exception. The way that ineptitude tips into oppression using tech is illustrated by the story of Robert McDaniel, as reported by Matt Stroud.[1]

McDaniel lived a quiet life in Chicago's Austin neighborhood, on the West Side. One day in 2013, McDaniel heard a knock on his door. Four people, including two Chicago police officers, stood outside. They told McDaniel he'd been identified as someone at risk for being involved in a shooting. The police officials had used a computational model to make this prediction. They didn't know if he was likely to be the shooter or be shot; the model didn't say. It just said that based on the geography and a few other demographic factors, the person living at this address was likely to be involved in a shooting.

The police wanted to talk, they said. Because they didn't know if McDaniel would commit a crime in the future or be a victim of crime, they were trying to cover all the bases. They had brought along a neighbor and a social worker. The social worker offered to help McDaniel get assistance with finding a job or accessing mental health services. The police backed this carrot up with a stick: they were watching him, they said. If he messed up or broke the law, watch out.

McDaniel was bewildered. He was just trying to get by; he had had a little interaction with the police regarding street gambling and minor marijuana-related issues, but he mostly kept a low profile. He was not interested in the model's conclusions, and told the visitors so.

The police kept coming back, though. They checked on him at the bodega where he worked. The social worker offered to get him into violence prevention programs and job training programs. He declined. But the police kept coming, with their offers. People in the neighborhood started to get suspicious. They saw the police hanging around McDaniel, visiting over and over, and they started to wonder: Was McDaniel a snitch? Rumors spread. One night, in a dark alley, McDaniel was shot. Some locals thought he was a police informant and they shot him for it. He recovered. The police came by to investigate. McDaniel told them to go away. Unfortunately, his reputation as an informant persisted, because everyone noticed each time the police showed up at his house. A month or two later, it happened again: someone *else* shot McDaniel for being a snitch, in August 2020.

McDaniel didn't know what to do. He told the police to go away. He explained that they were putting him in danger because of their presence. However, the police continued with their unwelcome attention, likely because the bureaucracy and the leadership of the Chicago police department were organized around believing the model and acting as if the model's conclusions were valid. But the thing that actually *caused* McDaniel to get shot was being tagged by the computational model and being targeted in the name of a misguided notion of safety.

Getting shot because of a computational model is nobody's idea of safety.

McDaniel was a victim of predictive policing, also called precision policing, a strategy that tries to use statistics to predict future crimes. Predictive policing comes in two flavors: person-based and place-based. Person-based policing, which was used on McDaniel, uses identity profiles. Computer models create an identity profile of who will commit a crime or be a victim of a crime in the future, based on specific crimes that have happened in the past. Local police then identify individuals or groups in the community who match the identity profile and pay them lots of (usually unwanted) attention. Place-based policing tries to forecast the place and time of a possible future crime. The system will identify an area, such as a 500-square-foot quadrant on a map, where crime is likely to occur based on past crimes. Police will often be dispatched to wait around in that specific area to address whatever crime happens where the computer says.

Both of these methods are largely ineffective.[2]

Predictive policing comes from the "broken windows" era of policing and is usually credited to William Bratton, former New York City police commissioner and LAPD chief. As NYC police commissioner, Bratton launched CompStat, which is perhaps the best-known example of data-driven policing because it appeared as an antagonist called "Comstat" on season three of HBO's *The Wire*.[3] "CompStat, a management model linking crime and enforcement statistics, is multifaceted: it serves as a crime control strategy, a personnel performance and accountability metric, and a resource management tool," writes sociologist Sarah Brayne in her book *Predict and Surveil*.[4] "Crime data is collected in real time, then mapped and analyzed in preparation for weekly crime control strategy meetings between police executives and precinct commanders." Comp-Stat was widely adopted by police forces in major American cities in the 1990s and 2000s. By relying heavily on crime statistics as a performance metric, the CompStat era trained police and bureaucrats to prioritize quantification over accountability. Additionally, the weekly meetings about crime statistics served as rituals of quantification that led the participants to believe in the numbers in a way that created collective solidarity and fostered what organizational behaviorists Melissa Mazmanian and Christine Beckman call "an underlying belief in the objective authority of numbers to motivate action, assess success, and drive continuous organizational growth."[5] In other words: technochauvinism became the culture inside departments that adopted CompStat and other such systems. Organizational processes and controls became oriented around numbers that were believed to be "objective" and "neutral." This paved the way for the adoption of AI and computer models to intensify policing—and intensify surveillance and harassment in communities that were already overpoliced.

Computer models are only the latest trend in a long history of people imagining that software applied to crime will make us safer. In *Black Software*, Charlton McIlwain traced the history of police imagining that software equals salvation as far back as the 1960s, the dawn of the computational era. Back then, Thomas J. Watson, Jr., the head of IBM, was trying to popularize computers so more people would buy them. Watson had also committed (financially and existentially) to the War on Poverty declared by President Lyndon Johnson upon his election in 1964. "Watson searched for opportunities to be relevant," McIlwain writes. "He said he wanted to help address the social ills that plagued society, particularly the plight of America's urban poor. . . . He didn't know what he was doing."[6] Watson wanted to sell computers and software, so he offered his company's computational expertise for an area that he knew nothing about, in order to solve a social problem that he didn't understand using tools that the social problem experts didn't understand. He succeeded, and it set up a dynamic between Big Tech and policing that still persists. Software firms like Palantir, Clearview AI, and PredPol create biased targeting software that they label "predictive policing," as if it were a positive innovation. They convince police departments to spend taxpayer dollars on biased software that ends up making citizens' lives worse. In the previous chapter, we saw how facial recognition technology leads police to persecute innocent people *after* a crime has been committed. Predictive policing technology leads police to pursue innocent people *before* a crime even takes place.

It's trIcky to write about specific policing software because what Chicago's police department does is not exactly the same as what LAPD or NYPD does. It is hard to say exactly what is happening in each police agency because the technology is changing constantly and is being deployed in different ways. The exact specifications tend to be buried in vendor contracts. Even if a police department buys software, it is not necessarily being used, nor is it being used in precisely the way it was intended. Context matters, and so does the exact implementation of technology, as well as the people who use it. Consider license plate readers, which are used to collect tolls or to conduct surveillance. Automated license plate readers used by a state transportation authority to automatically collect tolls is probably an acceptable use of AI and automated license plate reader technology—if the data is not stored for a long time. The same license plate reader tech used by police as part of dragnet surveillance, with data stored indefinitely, is problematic.

Every time the public has become aware of some predictive polic- ing measure, controversy has erupted. Consider the person-based predictive policing enacted by the Pasco County Sheriff's Office in Florida, which created a watchlist of people it considered future criminals. *Tampa Bay Times* reporters Kathleen McGrory and Neil Bedi won a Pulitzer for their story about how the Pasco County Sheriff's Office generated lists of people it considered likely to break the law. The list was compiled by using data on arrest histories and unspecified intelligence, coupled with arbitrary decisions by police analysts. The sheriff's department sent deputies to monitor and harass the people on the watchlist. Often, the deputies lacked prob- able cause, search warrants, or evidence of a crime. In five years, almost 1,000 people were caught up in the systematic harassment labeled "Intelligence-Led Policing." Notably, a large percentage of the people on the watchlist were BIPOC.[7]

The Pasco program started in 2011, shortly after Chris Nocco took office as sheriff. Nocco came up with the idea to "reform" the department with data-driven initiatives. "For 10 years, nobody really understood how this worked, and the public wasn't aware of what was going on," said Bedi, explaining the reporting project.[8] The sheriff built a "controversial data-driven approach to policing. He also built a wide circle of powerful friends," including local and national politicians, who didn't question his actions.

The harassment didn't stop there, however. Separately, the Sheriff's Office created a list of schoolchildren it considered likely to become future criminals. The office gathered data from local schools, including protected information like children's grades, school attendance records, and child welfare histories. Parents and teachers were not told that children were designated as future criminals, nor did they understand that the students' private data was being weaponized. The school system's superintendent initially didn't realize the police had access to student data, said Kathleen McGrory.

Once the investigation was published, civil liberties groups denounced the intelligence programs. Thirty groups formed a coalition to protest, and four of the targeted people brought lawsuits against the agency. Two bills were proposed to prevent this kind of invasion and misuse in the future. The federal Department of Education opened an investigation into the data sharing between the Sheriff's Office and the local school district. Fortunately, as a result, police analysts will no longer have access to student grades.

Many people imagine that using more technology will make things "fairer." This is behind the idea of using machines instead of judges, an idea that surfaces periodically among lawyers and computer scientists. We see it in the adoption of body-worn cameras, an initiative that has been growing since LAPD officers brutally assaulted Rodney King in 1991 and the attack was captured on a home camcorder. There's an imaginary world where everything is captured on video, there are perfectly fair and objective algorithms that adjudicate what happens in the video feed, facial recognition identifies bad actors, and the heroic police officers go in and save the day and capture the bad guys. This fantasy is taken to its logical conclusion in the film *Minority Report*, where Tom Cruise plays a police officer who arrests people before they commit crimes, on the recommendation of some teenagers with precognition who are held captive in a swimming pool full of goo. "It's just like *Minority Report*," a police officer marveled to sociologist Sarah Brayne, when the two were discussing Palantir's policing software.[9]

What makes this situation additionally difficult is the fact that many of the people involved in the chain are not malevolent. For example, my cousin, who is white, was a state police officer for years. He's wonderful and kind and honest and upstanding and exactly the person I would call on if I were in trouble. He and his family are very dear to me and I to them. I believe in the law, and I believe in law enforcement in the abstract, in the way that many people do when they have the privilege of not interacting with or being targeted by law enforcement or the courts.

But the origins of policing are problematic for Black people like me, and the frequency of egregious abuses by police is out of control in today's United States. Police technology and machine fairness are the reasons why we need to pause and fix the human system before implementing any kind of digital system in policing.

The current system of policing in the United States, with the Fraternal Order of Police and the uniforms and so on, began in South Carolina. Specifically, it emerged in the 1700s in Charleston, South Carolina, as a slave patrol.[10] "It was quite literally a professional force of white free people who came together to maintain social control of black, enslaved people living inside the city of Charleston," said ACLU Policing Policy Director Paige Fernandez in a 2021 podcast. "They came together for the sole purpose of ensuring that enslaved black people did not organize and revolt and push back on slavery. That is the first example of a modern police department in the United States."[11] In her book *Dark Matters: Surveillance of Blackness*, scholar Simone Brown connects modern surveillance of Black bodies to chattel slavery via lantern laws, which were eighteenth-century laws in New York City requiring Black or mixed-race people to carry a lantern if out at night unaccompanied by a white person.[12] Scholar Josh Scannell sees lantern laws as the precedent for today's policy of police using floodlights to illuminate high-crime areas all night long. People who live in heavily policed neighborhoods never get the peaceful cloak of darkness, as floodlights make it artificially light all night long and the loud drone of the generators for the lights makes the neighborhood noisier.[13]

The ACLU's Fernandez draws a line from slave patrols maintaining control over Black people to the development of police departments to the implementation of Jim Crow–era rules and laws to police enforcing segregation during the civil rights era and inciting violence against peaceful protestors to escalating police violence against Black and Brown people and leading to the Black Lives Matter movement. Fernandez points out that the police tear-gassed and pepper-sprayed peaceful protestors in the summer of 2020, fired rubber bullets at protestors, charged at protestors, and used techniques like kettling to corner protestors into closed spaces where violence could be inflicted more easily.

The statistics paint a grim picture. "Black people are 3.5 times more likely than white people to be killed by police when Blacks are not attacking or do not have a weapon. George Floyd is an example," writes sociologist Rashawn Ray in a 2020 Brookings Institute policy brief about police accountability.[14] "Black teenagers are 21 times more likely than white teenagers to be killed by police. That's Tamir Rice and Antwon Rose. A Black person is killed about every 40 hours in the United States. That's Jonathan Ferrell and Korryn Gaines. One out of every one thousand Black men can expect to be killed by police violence over the life course. This is Tamir Rice and Philando Castile." When Derek Chauvin, the police officer who killed George Floyd, was found guilty, it was remarkable because police are so rarely held accountable for violence against Black and Brown bodies.

Reform is needed. That reform, however, will not be found in machines.

Many academics are trying to grapple with these issues. Academic work is divided into specialties inside disciplines, and also centers around annual (or semi-annual) conferences. One important conference where interdisciplinary work on computational fairness can be found is called the Fairness Accountability and Transparency conference, or ACM FAccT. I have served on the program committee for the conference for several years, and every year we see a handful of papers that are trying to make "better" recidivism algorithms to predict which people are likely to be arrested again in the near future. This is likely because the field of algorithmic fairness started with Julia Angwin's groundbreaking ProPublica investigation of the COMPAS recidivism algorithm, which revealed algorithms can discriminate.[15] Since then, people have come up with many, many strategies for "improving" the algorithms, even though fairness scholar Jon Kleinberg definitively proved that it is mathematically impossible for the COMPAS algorithm to treat white and Black people fairly.[16] The way to make the world better is not to make better recidivism algorithms; it is to stop using these types of algorithms altogether.

Policing stats tell us number of arrests, not number of crimes. Black men are six times more likely than white men to be in prison in the United States, according to advocacy group The Sentencing Project. This is because of many factors, among them racist policies and racial disparities in enforcement of the war on drugs in Black communities for the past fifty years. If you feed this data into a computer, the computer model will conclude that Black people are more dangerous. "Decisions made by the system influence the data that is fed to it in the future," write Danielle Ensign and colleagues in a paper called "Runaway Feedback Loops in Predictive Policing."[17] They continue: "For example, once a decision has been made to patrol a certain neighborhood, crime discovered in that neighborhood will be fed into the training apparatus for the next round of decision-making."

The thing is, everyone is a criminal to some extent because everyone has done things that violate the law. For example, white and Black people use drugs and deal drugs at equal rates. Bias determines who gets constructed as a criminal; not everyone gets caught, not everyone gets punished, and some people get punished more than others. The unequal application of justice can be seen in crime maps. Look at a crime map for any major city, and it's pretty much the same as the map of where Black people live. Again, not because Black people commit more crimes, but because the things we call "crime maps" are actually arrest maps, and Black people are arrested for crimes at a higher rate. When you train algorithms on crime data, you are training the algorithm to over-police certain zip codes or geographic areas, because that is what has happened in real life in the past. You are training the algorithms to be biased.

Many do-gooders and criminologists fixate on crime statistics and commit the logical fallacy of assuming that crime data reflects the entirety of a problem.[18] This leads to software like HunchLab, developed by a certified B corporation (or B-corp) in Philadelphia called Azavea. A B-corp is a sort of hybrid between a for-profit and a nonprofit corporation. It's supposed to be a company oriented toward social good. And indeed, in many of its projects, Azavea is oriented toward the social good. Its Open Tree Map is a very nice open-source map of all the street trees in the city. But HunchLab, Azavea's crime prediction software, is a nightmare. HunchLab takes data on crimes, maps it geographically, and predicts the likelihood that a similar crime will happen again in that specific area. The idea was that it could be used to direct patrols to certain areas and reduce crime by having a greater police presence. A promotional screen-shot from the project shows a city map overlaid with different-colored squares. Each square indicates a location where a crime is likely to occur, and the color indicates the type of likely crime. The types include larceny, vandalism, burglary, auto theft, and robbery.

The HunchLab creators took data on past crimes, looked at the geospatial coordinates where the crime was reported, and generated a prediction of whether that crime was likely to happen again in that same spot. They then sold the software to the Philadelphia police, and to police departments in other cities, telling them that the software could be used to decide where to send patrols. This is typical of how other policing software like Crimespotter or PredPol or CompStat works.

Tawana Petty, the National Organizing Director for Data for Black Lives, summarized the problem with this kind of surveillance culture: "Pretending a thing creates safety, pouring millions of dollars into building out, and enforcing said thing, and pushing a media campaign that consistently calls the unsafe thing safety . . . actually makes the community less safe."[19] Robert McDaniel, who was algorithmically targeted by the Chicago police, might agree.

An art project called the White Collar Crime Risk Zones shows very effectively how the HunchLab approach is biased. Its creators mapped the location of white-collar crimes in New York City, which are quite prevalent but rarely prosecuted. The hotspots were in lower Manhattan. Wall Street, the global financial center, was lit up bright red. The Bronx, a largely poor borough where most police hotspots are identified, was largely free of white-collar crime predictions.[20] This is exactly the inverse of the predictions generated by something like HunchLab.

In a white paper accompanying the interactive map, the authors write: "In this paper we have presented our state-of-the-art model for predicting financial crime. By incorporating public data sources with a random forest classifier, we are able to achieve 90.12% predictive accuracy. We are confident that our model matches or exceeds industry standards for predictive policing tools." This is deadpan mimicry of the dry language of computer science papers. They also imitate the ambiguous, suspicious language of police modeling: "Our current model relies solely on geospatial information. It does not consider other factors which may provide additional information about the likelihood of financial criminal activity. Crucially, our model only provides an estimate of white-collar crimes for a particular region. It does not go so far as to identify which individuals within a particular region are likely to commit the financial crime. That is, all entities within high risk zones are treated as uniformly suspicious."[21]

The other major contribution of this paper is a composite image of the predicted white-collar criminal, assembled by downloading the pictures of 7,000 corporate executives whose LinkedIn profiles suggest they work for financial organizations. The authors averaged the faces of the executives "to produce generalized white-collar criminal subjects unique to each high-risk zone. Future efforts will allow us to predict white collar criminality through real-time facial analysis." The composite photo of the "white collar criminal" shows a smiling young white man.

Looking at the two crime mapping programs together shows us the way that a system of racialized logic guides the assumptions about police technology. The areas where white-collar crime is predicted in Manhattan are the areas where white people live and work. By extension, the Bronx neighborhoods where Black and Brown people live are the places where "other," more violent crime happens. In the words of University of Pennsylvania scholar Ezekiel Dixon-Roman, it is not only racialized logic, it is also extractive capitalist logic. Dixon-Roman, in a talk given to the NYU Institute for Public Interest Technology, said that technopolitical, sociotechnical systems target particular spaces and go after particular bodies. They impose their systems of racialized logic in order to feed their particular epistemologies of power. If we look at the computer as a greedy, racist, eugenicist rather than a digital savior, it changes the frame and helps to question the value of doing a thing with technology at all.

Consider a white-collar crime like tax evasion. The US Internal Revenue Service is underfunded. ProPublica reporting suggests that around $50 billion in tax revenue is lost annually because the IRS doesn't have the capacity to go after rich people for tax evasion.[22] The ultra-wealthy are skilled at tax avoidance. Multibillionaire Jeff Bezos, currently the world's richest man, did not pay any federal income taxes in 2007 or 2011. Other billionaires like Elon Musk, Mike Bloomberg, Carl Icahn, and George Soros have also managed to pay zero in federal income tax in recent years.[23] Instead, the IRS goes after poor people, who generally don't have the resources to hire expensive lawyers and accountants to deal with audits. Consider that against the price of paying for police brutality, which is punishingly expensive and comes at taxpayer expense. Sociologist Rashawn Ray writes: "Since 2010, St. Louis has paid over $33 million and Baltimore was found liable for about $50 million for police misconduct. Over the past 20 years, Chicago spent over $650 million on police misconduct cases. In the one year period from July 2017 through June 2018, New York City paid out $230 million in about 6,500 misconduct cases. What if this money was used for education and work infrastructure? Research suggests that crime would decrease."[24]

Thinking about the White Collar Crime Detector reminds me of the difference between what happened when I got pulled over with my Black father versus what happened when I got pulled over with my white husband. My dad and I were driving in a Lexus with a Harvard sticker on the back window, which he had put on because he was proud that his daughters were alums but also because signaling an Ivy League affiliation made him less likely to get pulled over as a Black man driving a nice car. My husband and I were driving in a Nissan Rogue, which cost half as much. We had the same veteran-focused car insurance, USAA, because of my father's military service.

With my father, the cop took his time running the plates. He walked slowly around the car, examining it suspiciously. He wrote up a ticket for having an expired inspection sticker (true) and gave a lecture about the importance of having one's paperwork up to date. We were lucky he was only giving us a ticket for an expired inspection sticker, he implied. He suggested that he was being benevolent, that we were committing a crime by crossing a bridge in our car and by existing at all. My father, who pretended to be polite through the entire interaction in order to avoid the situation escalating, lost his cool entirely after we drove away. He was upset for days afterward.

With my husband, the experience was different. The cop pulled us over on the Taconic State Parkway in Connecticut, took our insurance, and asked if we had a family member in the military. Yes, I said, my father had been a captain in the army. My husband chimed in that his father had also been in the army. "Thank them for their service," the officer said. "Consider this a warning. Drive slower." He handed back the insurance and registration, and went along his way. The car's inspection, by the way, was a month out of date.

Every Black and Brown person has a whole set of experiences like this, whether it's driving while Black or stop-and-frisk harassment. The ACLU of New York revealed that innocent New Yorkers have been subjected to police stops and street interrogations more than 5 million times since 2002. Black and Latinx communities have been the most targeted, and nine out of ten people stopped have been innocent.[25] These experiences make us feel vulnerable, showing us how thin the membrane of the civilized world is. We are only one step away from the crazies in the white hoods burning crosses in those moments. W. E. B. DuBois wrote about the double consciousness of Black folk, the experience of being inside one's own body while feeling racialized by the scrutiny of the white gaze. The difference between what happened with my Black father and with my white husband reminds me who is given a pass, and who is put into the system in America. White people in America don't have these kinds of moments. It's part of what makes many white people oblivious to racism in the world, and also to racism in tech. Tech is real life. We need to stop pretending that it isn't, and we need to talk about how technology gets deployed and against whom.

"Algorithms have no place in policing," said Hamid Khan, founder of Stop LAPD Spying Coalition. "I think it's crucial that we understand that there are lives at stake. This language of location-based policing is by itself a proxy for racism. They're not there to police potholes and trees. They are there to police people in the location. So location gets criminalized, people get criminalized, and it's only a few seconds away before the gun comes out and somebody gets shot and killed."[26]

There is an argument that crime has gone down because the police are using technology. This is not true. Crime is down overall. In Pasco County, Florida, where the Sheriff's Office was using predictive policing to harass citizens and schoolchildren, the Sheriff's

Office responded to criticism of its "intelligence-led policing" program by offering that burglaries and petty crimes were down during the course of the program. Journalists Bedi and McGrory looked closer at this claim and found that indeed this was a real trend, but it mirrored the decline in other surrounding counties that were *not* using the same strategies. Petty crime was down overall, and the Pasco decline was part of the larger trend. In fact, violent crime had gone *up* in Pasco during the time the harassment policies were in effect.

When you look at the amount of money that has been wasted on ineffective policing technology, it is horrifying. The Oakland Police Department in California has spent millions on license plate reader technology since 2006, and it scans license plates all over the city. In one absurd development, they had to implement a six-month data retention policy because they filled up an 80 GB hard drive with license plate reader data and didn't have enough money in the budget to buy more storage.[27] All of that collected data is not particularly useful. Only 0.1 percent of the license plates scanned match California Department of Justice's list of wanted or stolen cars. In 2019, Oakland police scanned 7.9 million plates. All of that data was accessible to the FBI and to other agencies on request for collaboration through a program called the Violent Gang Safe Streets Task Force. The Oakland Police Department's data was requested only twenty-one times.[28]

This unnecessary surveillance wasted countless computational resources and burned through energy. This wastefulness in police technology is contributing to the climate crisis as well. AI is not environmentally responsible: training AI models uses enormous amounts of energy, much of which is generated using fossil fuels, which contributes to global warming.[29]

Police have been militarized through a program of police departments purchasing or being gifted military surplus. On the one hand, it's good to reuse. On the other hand, military-grade weapons are seldom needed for routine policing. The reuse policy leads to things like NYPD officers holding machine guns in the subway stations while kids are commuting to school, leading to a skewed perception of danger. The rush to buy the newest tech is motivated by techno-chauvinism and leads to police departments buying useless things like a $150,000 robotic dog that is not merely absurd but also won't work in the rain or snow.[30] Biometric data, like facial recognition or iris scans or fingerprint scans, can easily be weaponized in the wrong hands. When the government of Afghanistan fell to the Taliban in 2021, there was a mad scramble to delete online data. People worried that their online histories and their biometric data would be used against them by the new regime.[31]

Just as the origins of police violence today can be found in the violence of enslavers and vigilantes, the data violence of machine learning can be found in its origins, the field of statistics. Some of the most venerated thinkers and originators of today's statistics are Francis Galton, Karl Pearson, and Ronald Fisher. These forefathers of statistics were responsible for core concepts like the Pearson correlation coefficient and Fisher information. The trio were also outspoken racists and eugenicists who inspired Hitler. They may have done good mathematical work, but their social thinking is reprehensible. It's possible to disentangle the art from the artist, and so too in math it is important to disentangle the mathematician from their racist, nonmathematical views. However, in this case, as with Linnaeus and taxonomy, it is worth looking closely at the way that statistics have been weaponized to promote racist ideas.[32]

"Current statistical methodologies were developed as part of the eugenics movement and continue to reflect the racist ideologies that gave rise to them," writes University of Pennsylvania sociologist Tukufu Zuberi in his book *Thicker Than Blood: How Racial Statistics Lie*.[33] He argues that statistical analysis must be deracialized and that this deracialization "[m]ust be part of a process that recognizes the importance of history and the goal of achieving racial justice for all . . . Studies relying on assumptions that impose a decontextual-ized racial identity in a social stratum should be replaced by better studies that incorporate more accurate assumptions. Social science must strive to develop a true human science where the dimensions of all population experiences are investigated within the broader social context in which we find ourselves."

The broader social context includes massive inequality and dis-crimination. It would be preferable to take this into account and design machine learning systems to rectify or subvert existing social problems. Saying that "digital is the future" and uncritically assum-ing a future guided by machine learning is equivalent to saying that white supremacy is the future, because the mainstream sociopolitical subtext of "digital" is a specific kind of capitalist white supremacy.

This coded language shows up everywhere once you are attuned to it. Consider this IBM AI governance report, which reads: "Exten-sive evidence has shown that AI can embed human and societal biases and deploy them at scale. Many experts are now saying that unwanted bias might be the major barrier that prevents AI from reaching its full potential. . . . So how do we ensure that automated decisions are less biased than human decision-making?"[34] This is problematic because it assumes that AI's "full potential" is even pos-sible, which has no evidence aside from the imagination of a small, homogenous group of people who have been consistently wrong about predicting the future and who have not sufficiently factored in structural inequality. The question of "How do we ensure that automated decisions are less biased?" reinforces this problematic assumption, implicitly asserting for the reader that computational decisions are less biased. This is not true, and IBM and other firms should stop writing things that include this assumption. The tech-nochauvinist binary thinking of either computers or humans is the problem: neither alone will deliver us. Sareeta Amrute encapsulates the problem thusly: "Holding onto the human as uniquely blame-worthy will only reinforce the utopian dream of elevating a class of experts above the raced, classed and gendered digital workers scattered across the globe at the same time that this dream's endpoint imagines transcending the human fallibility that causes ethical failures in the first place."[35] We need human checks on computational decisions, we need computational checks on human decisions, and we need additional safety nets plus the flexibility to change and adapt toward a better world. Fully automated systems are not the entire answer, and neither are human-in-the-loop systems. In the case of Robert McDaniel, the victim of predictive policing in Chicago, humans did review the computational prediction that he would be involved in a crime, and they did the follow-up that they thought was appropriate. However, their bureaucratic imperatives and their savior complexes meant they didn't listen to McDaniel when he said he didn't want their "help." He was ultimately hurt because of the computational and human interventions. What if the social worker had listened to McDaniel and convinced the police to stop visiting? What if the police had communicated more effectively with community members and realized that marked police cars showing up regularly at someone's home signaled risk, not public safety, in this context? In this case, as in so many others, a better resolution would have been to look more closely at the situation, examine assumptions, and chart a new path forward that was slightly different from the historical and computational precedent. This resolution would not be achieved by adding more data to the computational system—in fact, it could not be envisioned by a computational system at all. Computers aren't creative in the way that humans are. They don't have empathy, and they can't come up with alternative solutions or future scenarios as flexibly as humans (at their best) can.

Technochauvinists cloak their opinions in claims of "science" and "data." But as data scientist Cathy O'Neil says in *Weapons of Math Destruction*, algorithms are opinions implemented in code.[36] "Intelligence-led policing" is not data science. It's not scientific. It is just an oppressive strategy that upholds white supremacy. It's also similar to strategies used by Pearson and other eugenicists. "Maintaining the veneer of objectivity was crucial," writes mathematics researcher Aubrey Clayton.

> Pearson claimed he was merely using statistics to reveal fundamental truths about people, as unquestionable as the law of gravity. He instructed his university students, "Social facts are capable of measurement and thus of mathematical treatment, their empire must not be usurped by talk dominating reason, by passion displacing truth, by active ignorance crushing enlightenment." . . . In Pearson's view, it was only by allowing the numbers to tell their own story that we could see these truths for what they were. If anyone objected to Pearson's conclusions, for example that genocide and race wars were instruments of progress, they were arguing against cold, hard logic and allowing passion to displace truth.[37]

If you find yourself using a rationalization beloved of eugenicists in order to rationalize oppression, think again. This is a pretty good indicator that you are not on the side of the angels.

Eugenic echoes and deep historical biases are not limited to the algorithms used in the justice system. The next chapter illustrates the ways algorithmic bias and discrimination arise in another key area where people interact with the state: education.

5
Real Students, Imaginary Grades

Isabel Castañeda was excited to see her International Baccalaureate (IB) results, due to be emailed in the summer of 2020. Her Colorado high school, Westminster, had shut down for the COVID-19 pandemic in March and she hadn't seen her school friends in what felt like ages, but her IB scores were something to look forward to. Isabel was an excellent student, earning straight As in all of her classes. She was fluent in both English and Spanish, and had also studied French and Korean for years with the goal of someday becoming a translator. Everyone in her life predicted she'd do well on the IB tests.

The IB is an international diploma program based in Geneva that for many high school students in the United States promises an accelerated path to success via education. The IB offers a prestigious secondary diploma based on final exams in a range of subjects. IB classes are rigorous honors classes that end in standardized tests in specific subjects like math or US history. As with Advanced Placement (AP) tests, students who score high grades on IB exams can earn college credits at certain US universities. At Colorado State University, Castañeda's top choice college, high IB scores could earn her up to two years of college credit. For students like her, whose families struggle to imagine how they will afford the high sticker price for college, paying for two years of college instead of four was a golden ticket.

Castañeda was sure she'd do well. She had the grades, and her teachers were all optimistic about how she'd do on the tests. Historically, Westminster students tended not to score very high on the IB exams, but everyone knew that Isabel was going to be an exception. There was one problem: the in-person IB tests had been canceled because of the COVID-19 pandemic. The grades were to be assigned by algorithm.

Castañeda received the notification of her scores, checked online with great excitement, and then her heart sank. Her scores were far worse than anticipated. The score on the Spanish exam was particularly awful.

"I had previously taken AP Spanish and an IB class for native speakers, and I earned an A+ every semester both years," Castañeda told me. "My Spanish teacher and I believed that I would have gotten a seven, which is the maximum score. I personally thought I did pretty well on my internal assessments, on my spoken exam and on my writing exam. I even had people who grew up in in Spanish-speaking countries read over my written test because speaking fluent Spanish is very important to me. Everything was going very well; everybody believed that I was going to score very high, and then the scores came back and I didn't even score a passing grade."

She didn't score a passing grade. This is a young woman who spoke Spanish exclusively until first grade, who spends every summer speaking Spanish with relatives in Mexico, and who takes her language studies very seriously.

At first, Isabel wondered if she was the problem. "I thought, even though Spanish is my first language, maybe since I didn't grow up in a Spanish-speaking country it's possible that I could not do as well as people are testing in Mexico or Bolivia or whatever country," she said. "The confusing part, though, was we read the same kinds of books as the kids in Spanish-speaking countries." Before her school shut down for the pandemic, her Spanish literature class had been translating Albert Camus's novel *The Plague* from Spanish to English.

Then, she reread the email from IB about algorithmic grading and realized something bigger was happening. "It doesn't make sense that I could score the same as someone who speaks very broken Spanish," Isabel said. "There's nothing wrong with that, but for the purposes of testing, I think that I can speak the best Spanish possible in a country that doesn't have Spanish as its main language. So, even though I don't believe myself to be a 100 percent fluent speaker like the people who live in other countries, I still think that a non-passing evaluation that would be given to someone who can barely speak Spanish is not a fair evaluation."

Isabel was right. Her failing grade was not fair, and it didn't make sense.

The technocrats at International Baccalaureate headquarters in Geneva had chosen not to administer the usual in-person exams—an appropriate choice given that in the spring of 2020, every country was facing sky-high rates of infection, hospitalization, and death from COVID-19. In-person school was considered dangerous at the time, and other tests like the AP, SAT, and ACT were all canceled or administered remotely. Instead of canceling the IB exams and issuing refunds on the testing fees, which schools and students had already paid, IB decided to create an algorithm that would assign imaginary grades to real students. They would "predict" the students' grades using a computer instead of assessing the students' actual academic performance.

The process they used was a standard one in data science. IB personnel took all the available data on the students and their schools and fed it into a computer. The computer then constructed a model that outputted individual student grades. The shorthand for this process is "used an algorithm." The longer version is the process described in an earlier chapter of this book, but in this case the computer predicted grades rather than predicting who was likely to pay back a loan. The data scientists ran some tests, decided the results were within an acceptable margin of error, and sent out the grades that they claimed the students would have gotten if they had taken the standardized tests that didn't happen. It's a legitimate data science method, similar to the methods that predict which Netflix series you'll want to watch next or which deodorant you're likely to order from Amazon.

The real problem was the algorithm. Data science stinks at making predictions that are ethical or fair. The IB results at Westminster are a good example of why looking at the data alone is not a good proxy for predicting individual student achievement. Westminster students are predominantly Black and Brown, with many kids who are heritage Spanish speakers who grew up in the United States, rather than in a Spanish-speaking country. Additionally, 78 percent of Westminster students come from low-income families, and Greatschools.org rates the school as performing far below the state average in key measures of college and career readiness. If you were just looking at the data on the school overall, it would not paint a picture of a thriving academic environment.

There are multiple ways to look at the data. IB's own reports show that low-income students overall score lower on college readiness assessments than high-income students. IB is supposed to provide a ladder, a pathway for motivated students to succeed academically regardless of background. It generally works: low-income IB students tend to rise above their peers academically. In a 2015 report, IB found that more than three quarters of low-income IB students enrolled in college. Their six-year college completion rates were very good.[1]

IB's algorithmic grading program undermined the efforts of teachers, students, and administrators. "Everyone I know got downgraded one or two levels," Castañeda told me. "Our school has historically done badly on IBs, but the 2020 class was more dedicated and hardworking than in past years. It's not fair that our scores were brought down because of our school's history. It's unfair to punish smart people for where they live." Isabel had deliberately applied to a college that would award credit for high scores on IB and AP exams. If she did as well as her teachers expected on her IB exams, she could graduate college early, saving her family tens of thousands of dollars. Instead, like many other students who were downgraded, she felt that "[w]e're getting screwed over by IB because we wanted to save money on college."

Thousands of students and parents protested online and in person. The IB reported that 700 of its 3,020 schools around the world asked for student grades to be reviewed.[2] The bureaucrats who decided to use a computer to assign grades are guilty of technochauvinism. They thought that the algorithm would offer objective, fair grades and that the computational solution would be sufficient in a time of global crisis. It wasn't, because a grade in education is a social decision, not merely a mathematical one. Computers are rarely the fairest solution when it comes to social decisions.

Education itself is deeply unfair. In the United States, the country whose education system I know best, the single biggest determinant of educational success is income. Rich students do better in school than poor students, almost always and across the board. Eliminating poverty would be the best thing we could do to improve educational outcomes. But to a computer, there is no sting in the fact that students from affluent backgrounds do better than their under-resourced counterparts. The computer doesn't feel shame at the way the United States has failed the 10.5 percent of its citizens living at or below the poverty line. It simply predicts that the rich kids will earn higher grades than the poor kids, and then it waits patiently to fulfill its next command. Education is supposed to help students from all communities get ahead. Using algorithmically determined grades puts the lie to the entire American dream of education as a ladder to success.

Even if somehow we could make an algorithmic system that was fair to all, regardless of social or economic class, the system would likely discriminate on race. If we could solve for race, the system would likely be biased against some intersection of race, class, gender, social status, ability, or many other factors. People may be created equal, but privilege is not evenly distributed.

Economic inequality is inextricably connected to racial disparity. School districts that serve primarily students of color get $23 billion less in funding than districts that serve primarily white students, according to a 2019 report by the nonprofit EdBuild. Online learning during COVID-19 was better for white students than for nonwhite students. "Schools in Broward County with higher percentages of minority and/or low-income students and schools already struggling with academic achievement fared the worst as classes moved online," wrote The Markup in 2020 in its analysis of Florida school data. "In many cases, whiter, wealthier schools with higher achievement scores actually improved attendance after the switch."[3] A computer is not devastated to admit that school segregation persists in the United States, more than sixty years after *Brown v. Board of Education*. A computer is not roused to take action and rewrite the rules to remedy this profound injustice. The computer will just take in historical data and predict that Black and Brown students will do worse than white students at school.

One of the inputs to the IB algorithm was historical performance data for each school. At Isabel Castañeda's school, Westminster, the computer assumed that all the students (who are mostly low-income students of color) would continue to do poorly. Another input to the IB algorithm was teacher prediction of the students' grades. Teachers tend to predict lower grades and worse outcomes for Black and Brown students compared to white students. This bias is well known in the education community and ignored in the data science community. IB data scientists gave teacher predictions the most weight in their algorithmic model, allowing a very human bias to prevail in the computational system.

In Isabel Castañeda's case, we don't know exactly what her teachers predicted, but a few things are possible. Her teachers could have given her excellent evaluations, but the predictions for Westminster students overall might have prevailed in the model. Or the teachers might have given her what they thought were good scores, but the scores might not have been high enough to satisfy the algorithm. One person's measure of good isn't always the same as another's. Teachers with more privilege often know how to game the system to ensure that their students do better in complex ranking systems.

Strangely, it is not uncommon to assign imaginary grades to real students. Education data is filled with measurements that are statistically valid but nonsensical. For example, New York City schools are evaluated based on a School Quality Report, one dimension of which claims to evaluate how well individual students do compared to demographically similar students at different schools. "The Comparison Group results estimate how the students at the school would have performed if they had attended other schools throughout the city," reads the New York City Department of Education guide. "By comparing the school's results to the Comparison Group, a reader can assess the school's effectiveness at helping students improve and exceed expected outcomes."[4] In other words, each real student is compared to an imaginary group of hypothetical students at a different school in the same city.

This is absurd. Math is wonderful, but it cannot rewrite a person's history. It makes little sense to calculate how a student *would have done* if they had attended another school. Another problem with the notion of imaginary grades is the difference between aggregate outcomes of groups and the outcomes of individuals. We might say that, on average, men have more upper body strength than women, but that does not mean every individual man is stronger than every individual woman. Serena Williams is clearly stronger than her husband Alexis Ohanian, for example. And sometimes the stakes are low enough that generalities can still be effective. Crude generalizations work for Netflix predictions because the stakes are very low. If the Netflix algorithm suggests a show and I don't like it, I ignore it and move on with my day. In education, the stakes are much higher. Your school record follows you for years: when I was twenty-five and long out of college, I applied for a job that asked for my SAT scores.

It is true that over time, everything evens out. Flip a coin a million times, and you will indeed see it end up heads half the time and tails the other half of the time. But we don't flip a coin to decide people's fate. We want fate to be in our own hands, especially when it comes to educational fate, which holds the key to people's economic and personal success. This is another case of the mathematical colliding with the social. Lots of people grade on a curve, and sometimes the curve means that students do worse than they expected. I think that's nonsense. As an educator, I'm thrilled when all of my students earn As in my class. It means they have all mastered the material. There is room in the educational system for both kinds of instructors: professors who always penalize a few students as well as professors who help all their students thrive.[5]

It is well known that standardized tests historically favor wealthy, white students.[6] It is also well known that algorithms discriminate and reproduce existing inequalities. So, any algorithmic grading system for any standardized test is going to predict that students at wealthier, predominantly white schools will do better than students at under-resourced schools that serve primarily Black and Brown students. It's simple mathematical logic. One of the big misconceptions of data science is that it provides insights. It doesn't always. Sometimes the insights are merely things that the data scientists didn't know, but people in other disciplines already knew. There's an important distinction between what is unknown to the world versus what is simply unknown to you. Data scientists in general need to do more qualitative research, and talk to experts in relevant fields, before designing and implementing quantitative systems.

Instead, in any case where there is an algorithmic system in place for an educational decision, there should be a sensible appeal process that is responsive and human-staffed and has a reasonable chance of success. Isabel Castañeda should have been able to click a button to protest her grade, and IB should have assessed the evidence within hours and revised her grades upward. Even grocery stores have this feature: if you order eggplant parmesan and they deliver lasagna, you click a button or call a phone number and they issue a refund without a fuss. We know that algorithmic systems are going to fail and discriminate; we should be prepared to mitigate the shortcomings immediately.

Castañeda, fortunately, thought to protest her algorithmic grades. IB didn't change her grades, but she appealed the decision to Colorado State, where she had been admitted. As part of her appeal, she showed the university administrators the national media coverage of her case that explained the absurdity of algorithmic grades. The administrators agreed with her, granting her the college credits that she would have earned for her excellent high school performance. Isabel will likely be able to graduate in less than four years and begin her dream career as a translator, saving her and her family thousands of dollars in tuition and student loans.

Some may argue that an appeals system is likely to be more expensive and more elaborate, undermining any cost savings from the algorithmic system. Or people are going to abuse the appeals system if they don't like their grades. Both are valid concerns. It's more expensive and time-consuming to have a rapid appeals system in addition to an algorithmic system. It's also the only way to ensure a measure of justice. And it begs the question: If the algorithmic system is definitely going to discriminate, and people are going to have to manage the appeals process, why bother using an algorithmic system at all?

A better alternative when standardized tests fail is to reassess what the original purpose of the test was. This strategy was one adopted by several states for the bar exam, a system called diploma privilege. The bar exam is a long, complicated standardized test that students take after law school in order to qualify to practice law in a particular state in the United States. Each state's laws are different, so the bar exam is one of the ways that students demonstrate their proficiency with the state's laws as well as with lawyering in general. The bar exam is usually administered over multiple days, and some people take off months from work and life to prepare for it. This regular system is not great: it disadvantages the students who can't afford to study full-time instead of working, and students with disabilities have traditionally been required to fill out reams of paperwork documenting their disabilities and requesting accommodations. However, for years, this was the gatekeeping function for the profession.

When the COVID-19 pandemic hit, it was clear that the bar exam could not happen. Putting thousands of law students into poorly ventilated rooms to take a high-stakes exam for hours was a public health risk that nobody wanted to assume. Instead, several states adopted what they called diploma privilege, which allowed graduates of American Bar Association–accredited law schools to begin practicing immediately after graduation. Most people go to work for law firms and practice in collaboration with or under the supervision of more senior lawyers. The system worked extremely well.

What didn't work well was trying to replace in-person bar exams with remote exams. Washington State tried this. They tried to use ExamSoft, a remote proctoring service. It was a disaster. ExamSoft flagged as possible cheaters more than one-third of the 9,000 students taking the California exam in October 2020. In Michigan, they tried to use ExamSoft and the software crashed in the middle of the exam. In Pennsylvania, they tried to use it and students brought a lawsuit.[7]

ExamSoft and Proctorio and other remote proctoring services are part of a class of software known as edtech surveillance solutions.

Edtech surveillance systems are widely loathed by anyone forced to be monitored by them. They don't recognize people with dark skin, religious headwear, or facial features that diverge from the most typical shapes and patterns. They're just . . . bad software. For starters, if you are a person with darker skin, you often need to shine a bright light directly in your face and keep it there for the duration of the exam because the facial recognition software struggles to recognize your skin, especially in average or low light. The software requires users to stay put for long stretches. You can't get up to take a break or go to the bathroom or get a tissue or get a drink. Students with disabilities are not parsed well by these systems. If you're Blind, for example, the system doesn't think you're looking in the right spot. People who are not taking the exam in a quiet place get flagged as cheating. If your kid or a family member walks through the frame, the software flags you for cheating. If you blow your nose on a tissue, the system might register it as paper and flag you for cheating. The broadband-hogging, memory-intensive software discriminates against people without high-speed broadband access or expensive up-to-date computers. Broadband access is something that people take for granted in large cities, but access is not guaranteed outside of US metropolitan areas. On one tribal reservation in Washington State, for example, 80–90 percent of the residents lack high-speed internet, and cell phone dead zones abound.[8] The list of problems goes on and on. It's crucial that we address the problems with these edtech surveillance systems now, because the virtual schooling strategies forced by the pandemic have spawned a wave of enthusiasm for online proctoring and other digital methods that is unlikely to abate.

Chris Gilliard, a Macomb Community College professor and noted edtech skeptic, is pessimistic about the future of surveillance edtech. "It's not going to stop at students," he told me. "If you're surveilling students, the next step is to surveil instructors. I wish people could see that, because I think some of the hardliners who think it's okay to surveil the students would recoil if it were proposed to be done to them." Some of the surveillance edtech companies have explicitly said they plan to expand to instructor surveillance. Even the ones who haven't yet will likely do so in the future, Gilliard says, because of the market imperative for infinite growth and expansion.

One problem with adopting surveillance edtech is that schools get stuck in enterprise software contracts. A private individual can buy a monthly or yearly license for personal software. As of this writing, a one-year license to use Microsoft Office 365, a collection of software that includes Microsoft Word, PowerPoint, and Excel, costs $69.99 per year. Enterprise software licenses allow an organization to buy licenses in bulk for users across the enterprise, meaning that all the faculty and staff and students in a school will be on the same platform. It is usually cheaper to buy software licenses in bulk, just as it's usually cheaper to buy toilet paper in larger quantities.

Once you have gone through all the trouble of making an enterprise software contract, you're locked in. "At first there is optimism, but once places make that investment, they're very unlikely to go back and say they made a mistake," said Gilliard. That's why schools end up digging in on using software that clearly doesn't work as intended or doesn't benefit students. In fact, school software is a wasteland of neglected products. Most apps that schools buy are not used. "Schools will buy these licenses, and then they never really get touched," said Ryan Baker, director of the University of Pennsylvania Center for Learning Analytics. His 2018 analysis found that a median of 97.6 percent of app licenses were never used "intensively," meaning for ten hours or more, between assessments.[9]

Schools and colleges, like government in general, need to get with the program and figure out how to balance individual cloud software licenses as well as enterprise contracts. In addition, another possible solution is to put a morality clause into enterprise software contracts. Enterprise software contracts are negotiated by humans:

salespeople on the software side, procurement departments and lawyers on the school side. School districts, colleges, and universities could add a morality clause to their contracts that says if the software is found to discriminate or to violate the law, the contract can be dissolved. There could even be a clawback clause that says that the software company has to refund the fees paid if the school finds evidence of discrimination that is not remedied within, say, two weeks of discovery. Why don't we have this? Probably because of technochauvinism, and lawyers feeling like the software company has the power in the relationship because the software salespeople claim to understand the technology better than the customers.

It's important to act soon to stop the spread of biased edtech and algorithmic systems because these systems are being adopted right and left by clueless educational bureaucracies. There are going to be infinitely many technochauvinist calls to transform online education and use algorithmic tools that promise to evaluate or assess individual student learning as we go further into the era of big data. As badly as algorithmic grading systems and online proctoring systems work, people are likely going to try to implement them over and over in the next few years. Resist the siren call. These systems don't work as well as the people selling them to you might claim. As we'll see in the next chapter, relying on technology to do all the work can end up restricting access even when the intent is to make an environment inclusive.

6

Ability and Technology

Richard Dahan loved his job at the Apple Store in the Montgomery Mall in Bethesda, Maryland. He got a kick out of helping people choose new devices, the way they got excited about the slick packaging and the possibilities they saw in owning some new tech. Dahan started work as a specialist, then after a few years moved up to be an expert, doing team development, troubleshooting, and taking care of customer issues. His particular store, near his home in Frederick, Maryland, had a reputation in the Deaf community because Dahan, who is Deaf and fluent in American Sign Language (ASL), worked there. Being able to get tech support or shopping advice in ASL was a relief for many customers. "People told me they would drive two hours to my store, just to see me—because they could communicate with me in their own language," said Dahan.

To communicate with hearing customers, Apple provided him an iPad with the Notes app installed, to type back and forth. Things went well, except when they didn't. "For the older generation, it was hard to type on the iPad," said Dahan. "People who came in from other countries, other cultures—that was tough." Nevertheless, he thrived in the job for the first few years.

Apple has a stellar reputation for making products that serve a range of abilities. Apple's software is popular in the disabled community broadly, and products like its VoiceOver screen reader are considered to be some of the best available. The company is known for accessible design across multiple dimensions. Sign language interpreters are available in Apple stores by appointment in the United States, United Kingdom, Australia, Austria, Belgium, Canada, France, Germany, Italy, Sweden, and Switzerland. Assistive-Touch, a feature for Apple Watch, is designed for those with limited mobility, so a user can navigate the display with a pinch or clench instead of a tap. Background Sounds, a feature that adds background noise, is designed to help neurodiverse people to reduce distractions or manage noisy environments.

Apple is also known for its public commitment to inclusion. The company has a partnership with Gallaudet University, Dahan's alma mater, which serves the Deaf and hard of hearing communities. Apple recruits on Gallaudet's campus (that's how Dahan got his job) and encourages students and faculty to use the company's products. In 2018, the company launched a curriculum called Everyone Can Code, which specifically had accessible features built in, and launched the curriculum at eight schools for Blind and Deaf students. "Apple's mission is to make products as accessible as possible," said Tim Cook, Apple's CEO, in a press release connected to the curriculum launch. "We created Everyone Can Code because we believe all students deserve an opportunity to learn the language of technology. We hope to bring Everyone Can Code to even more schools around the world serving students with disabilities."[1]

Apple's retail customers, however, were not always committed to inclusivity, in Dahan's experience. "Some people came in and they were rude. They wouldn't type on the iPad with me, or wouldn't work with me because I was Deaf," said Dahan. "They would avoid me and ask to work with other staffers. There was a manager who enabled this behavior, who would say, 'It's the customer's right to decide who they work with.' I would think, 'If someone came into the store and they didn't want to work with a person of color, would that be acceptable?' It felt like they were oppressing me."

The dominant narrative is that technology enables accessibility, but Dahan's experience with Apple shows that the situation with tech and ability is nuanced. There's no one-size-fits-all approach to disability, so it's hard to make generalizations about disability and tech. I talked to many people with disabilities about their experiences with tech, and I learned that today's tech is marvelously empowering until it isn't. Once you reach the outer limits of the tech's capacity, it becomes marginalizing. Tech is great for accessibility, but there is still much to be done. The nondisabled community needs to listen better and not assume that because we can provide technology, all the problems go away.

"The issue with tech and the Deaf world is, there's no way to talk about it," said Dahan. People are so excited about the *possibilities* of tech that there's very little opportunity for nuanced discussion about the many advances that remain to be achieved. It's considered disloyal, somehow, to point out that if GPS doesn't work in a deep underground subway station, using Google Maps audio directions is not going to help a Blind person navigate their way onto a train. Apple, in particular, has an enormous fan base among people of all abilities. The fans' breathless enthusiasm and brand-loyal optimism tends to drown out conversations that suggest the latest i-whatever might not be the best, most exciting thing ever. A more measured approach, one that suggests that there are still great lengths to go before tech is truly inclusive, would be more honest and less technochauvinist.

Dahan's experience at Apple, which he shared publicly during the #AppleToo activism campaign, shows that accessibility is still a problem inside the trillion-dollar company. "I grew up Deaf, my parents are Deaf, and I went to a Deaf residential school. I grew up in a Deaf world," Dahan told me. "I identify strongly as a Deaf person and with Deaf culture. Apple was the first place I experienced discrimination and I ran up against barriers."

Dahan's manager, the one who didn't have his back with customers, gave him a bad review. "The performance review was unfair," said Dahan. "He suggested that my numbers were not up to par." The manager implied that Dahan was missing out on important information being discussed in the store's daily download meetings. Dahan was frustrated; he *knew* he was missing out on information, because the accommodation the manager had offered was inadequate. A team meeting, with lots of attendees in a cacophonous room, is a difficult situation to navigate if you're Deaf or hard of hearing and don't have an interpreter. People have a range of hearing abilities, and certain environments are more helpful than others. The design of Apple stores is particularly noisy: the company's aesthetic favors hard surfaces and high ceilings with few textiles or soft surfaces to dampen sound. Recommended architecture for Deaf and hard of hearing people is precisely the opposite. Acoustic ceiling tiles or sound-muffling floor coverings and underlayments can help filter out unnecessary noise and prevent the echoes that cause painful reverb for people who use hearing aids. At the Montgomery Mall Apple Store, one of the four walls is entirely made of glass. It looks great, but glass is generally inefficient at reducing sound transmission.

Ordinarily, a noisy meeting is the kind of situation where a Deaf person would use an interpreter. There are options: either a human interpreter who would come to the meeting and translate into ASL or a video interpreting service that would do the same. Neither option was made available to Dahan. Instead, the manager asked for a volunteer to type what was happening at the meeting so Dahan could read it. The volunteer, another employee who was also trying to listen and learn, would type three to five words every minute or so—hardly an adequate representation of what was happening in the meeting. Dahan asked for an ASL interpreter so he could be more involved with the meetings. "My hearing peers bene-fited from the incidental learning where they could listen and learn around them," he said. "I wanted the incidental learning experience as well." Another feature of the daily download meetings was when workers were given the opportunity to lead the discussion if they wanted to move up to become a manager. An interpreter, Dahan explained, would be a reasonable accommodation that would allow him to interact with his coworkers and customers. It would give him equal access to communication and information, which he needed in order to progress in his job. The store refused.

"I went to HR," said Dahan. "I tried to explain, 'I can learn from different conversations, which will make me better at my job, but I need to be able to hear what's happening. I can do that with an interpreter.'" Under the Americans with Disabilities Act (ADA), companies with more than fifteen employees are required to pro-vide an interpreter if requested by an employee who needs one.

Eventually, Apple did offer to provide Dahan a video interpreter. This technology, usually called video remote interpreting (VRI), has been around for a while. Using VRI means that a sign language speaker connects with an interpreter via video. The Deaf and the hearing per-son are in the same location, and the interpreter is remote. The inter-preter signs one side of the conversation and speaks the other side. This kind of video translation works well for some situations but is inadequate for others.

"VRI is okay for one-on-one meetings, but in a big meeting there is background noise and the Wi-Fi freezes," Dahan said. "It's better to hire a physical interpreter to come into the meeting. For some-one who's Deaf, having an interpreter in this kind of environment is like going from black-and-white to color television." He tried the video interpreter, but the technology failed. The interpreter, who was listening remotely through an iPad, had trouble hearing the speakers at the meeting. If many people were talking at the same time, their voices overlapped and the interpreter couldn't hear any of them. Background noise was a problem for the interpreter. Sometimes the video froze, or the person speaking was so far away from the iPad that the interpreter couldn't hear them. Dahan couldn't follow what was happening. "I asked repeatedly, but Apple gave me the runaround," he said. "They were trying to find different ways to avoid getting an interpreter. I feel like they actually think that their products and apps can cure my Deafness by trying to convince me that their products can be used as my job accommodation."

Dahan told me his story via a video relay service, which is similar to VRI except that the Deaf person, hearing person, and the interpreter are all in different locations. I used my telephone, and Dahan used his visual screen. It worked well, although the interpreter missed some technical concepts and mistranslated at least one major component of the story. Dahan and I addressed the error by sending the text of our interview back and forth for editing. I was grateful that the technology existed so that Dahan and I could communicate synchronously and asynchronously. At the same time, I could easily see how in-person communication would have been more efficient and information-rich. Talking about technology or any specialized field is always a challenge. Talking about it in two different languages, with an interpreter who didn't have the field's specialized vocabulary and was unfamiliar with the context, was an even bigger challenge. It was clear to me that an in-person interpreter would have been the best choice to allow Dahan to participate fully in the Apple meetings.

Another situation arose where Dahan wanted to be involved in a social event with his coworkers, a picnic that store workers had organized using company email. "I asked Apple to provide an interpreter so I could attend the event," Dahan said. "Apple said, 'No, it's outside of work.'" An ASL interpreter costs about $100 per hour, which is what Apple charges for a power adapter.[2] It is also a tax-deductible expense.

A few years later, the company launched a service called Sign Time, which provided on-demand sign language interpreters via video in the store. Though Apple stores in the United States had long had a phone support team for languages in addition to English, they had not had one for ASL. After the launch, there were complaints. The idea was that a customer could click on the app and connect with a sign language interpreter. "They encouraged employees to use Sign Time, but it does not work as well as VRI," said Dahan. "There were a lot of complaints that it didn't work because the store was loud. Apple said, 'Okay, we'll just use AirPods—the Deaf employee can give them to the customer if the iPad isn't loud enough.' If you think about it, that's not realistic. You can't just pass around used AirPods. Having an interpreter is more practical." Practical it might be—but it would not position Apple technology as a total accommodation, which would undermine the brand positioning.

Talking to Richard Dahan about the outer limits of assistive technology helped me think through the way that technology and ableism intersect. I had started thinking seriously about oppression at the intersection of technology and disability about ten years ago, when a Blind student signed up to take my then-brand-new data journalism class. By then, I had taught students with various disabilities in my classes over the years, and each time I happily learned about the necessary accommodations and incorporated modifications into my teaching. However, this particular time I was stumped. My data journalism class relied on three core technologies: Microsoft Excel, a spreadsheet program; Tableau, a data visualization program; and a SQL database program. Excel would likely be fine, because Microsoft was an established company with a history of making its products accessible. For SQL it would likely be fine to use a screen reader since most of the commands could be entered in plain text without a graphical user interface. Tableau, however, was the big problem. It was a data *visualization* program. You use it by finding data on the web and putting it into Tableau, which arranges the data into a variety of graphical formats. I didn't see any way to make it accessible to a Blind student. It seemed that the Tableau software designers didn't either: no accommodations were available through the program itself, so it was up to me.

"How do other people teach data visualization to Blind students?" I asked the consultants at my school's disability services center. They looked at me blankly. Apparently, this had never come up before. They weren't even sure what data visualization was. Not only had I had just started teaching the class, but at that time—the early 2010s—data journalism was still fairly new as an academic discipline. I was shocked to discover that even with all of the advances in teaching and technology and pedagogy, I had stumbled into unknown territory.

The student ended up having a schedule conflict, and decided to take a different class. That brief experience, though, made me realize that I had a lot to learn about making technology (and my classes) more accessible, both to Blind or low-vision students and to my students in general.

I had a chance to learn and grow the next semester, when another Blind student enrolled in a class I was teaching on journalism ethics. The class was a blend of lecture and discussion. I checked with the disability services center: What was the best practice for dealing with the slides I'd be using during lectures? My liaison said I should read the text on each slide and describe any images. I should create the lecture slides in PowerPoint because, at the time, Microsoft products tended to be the most accessible. I should also make sure the student had access to the slides.

I was surprised to find that I hadn't been reading the slide text aloud in lectures. I realized I was following advice I'd been given in a public speaking course in the late 1990s, that a speaker should never read their slides. Times had changed, and so would I. I started reading my slides aloud. Although it slowed down my lecture speed a bit, it became easier for all of the students to take notes.

I could have emailed the slides to the one student after every class, but instead I decided to make them available to everyone on Blackboard, the learning management system my school used. Blackboard worked reasonably well with screen readers, software that reads aloud what's on a computer screen. Instructors have different philosophies about distributing slides and lecture notes. I had previously been in the camp that didn't distribute slides, preferring for students to take their own notes in class as a pedagogical technique. Multiple studies have shown that writing notes by hand helps students gain deeper understanding of class material as compared to taking notes on a computer. This time, however, I posted the lecture slides on Blackboard before every class. I saw that students were downloading them, and presumably reading them. I was pleasantly surprised. I had been using my slides as cues to remember what to talk about in my lecture. The students were using them to engage with the class material a bit more—and they were still taking notes in class. I checked in with my student a few times during the semester to see if I needed to do anything differently. Things were going well, they said. I realized that the studies about student note-taking and cognitive effects probably didn't include students with disabilities in their samples. Using a Braille computer to take notes is obviously more helpful than taking notes by hand for a student who is Blind or has low vision.

Learning about accessibility, and integrating accessible teaching techniques into my practice, made me a better teacher. This is typical. People refer to it as the curb cut effect. Curb cuts are the small ramps cut into street corners. They were originally implemented to make it easier for wheelchair users to cross the street, but they ended up benefiting everyone. Curb cuts make it easier to navigate with a walker or a stroller. They are good for people with balance issues, who benefit from not tripping when they step on or off the curb. People with carts and dollies use the ramps to get from the street to the storefront. The effect of curb cuts, like the effect of accessibility in general, is to make things better for everyone—not only for people with disabilities.

As I learned more about accessible technology, I was relieved to discover that the way I'd started, by updating small practices, was considered an appropriate strastegy. Sarah L. Fossheim, a designer who focuses on accessibility, said on a panel: "People get overwhelmed when you frame accessibility as 'you have to be compliant and there are all of these checklists.' Most of the people I talk to completely freeze and end up making *nothing* more accessible because they want to be fully compliant from the get go." Fossheim recommends beginning where people are. "I try to say, take a step back and don't try to aim for the highest," she said. "Let's just try to improve what you have, and in the next iteration we'll improve a bit more, and we'll experiment in between. Then, suddenly, things start improving much faster."[3]

The way you design for digital accessibility in teaching is similar to the way you design for digital accessibility in any other realm. Checklists help. A "508 checklist" is the federal government's checklist for making digital content accessible. At the US Department of Health and Human Services, the checklist for creating a web page includes items like "Synchronized captions are provided for non-live audio (YouTube videos, etc)" or "All images, form image buttons, and image map hot spots have appropriate, concise alternative text." These are good web design practices, but not everyone provides alt text for images or inserts video captions unless reminded to or required to do so. 508 is a shorthand reference to Section 508 of the Rehabilitation Act of 1973, amended by Congress in 1998, which is about inclusion in the digital sphere. Specifically, this section requires the federal government to provide people with and without disabilities comparable access to electronic and information technology. Interestingly, national security systems are not required to be accessible.[4] Although private companies are not always required to adhere to 508, the checklist is useful for understanding ADA compliance. Compliance checklists may seem overwhelming at first, but they are an essential tool in making digital technologies accessible.

Accessibility mandates have gained teeth in recent years because of lawsuits levied against individual websites and platforms that do not make their content accessible. Accessibility advocates usually use a carrot and stick approach. One notable case began in 2014 when the National Federation of the Blind (NFB) sent a letter to Scribd, a content platform, alerting them that their over 40 million electronic books and documents were not accessible using screen readers. NFB invited Scribd to collaborate on making the material accessible. No response. So, NFB sued Scribd on the grounds that access was mandated under the ADA. The Americans with Disabilities Act, signed into law in 1990 by George H. W. Bush, assures equality of opportunity, full participation, independent living, and economic self-sufficiency for individuals with disabilities. It states that "physical or mental disabilities in no way diminish a person's right to fully participate in all aspects of society, yet many people with physical or mental disabilities have been precluded from doing so because of discrimination" and seeks to address discrimination as well as social, vocational, economic, and educational disadvantages encountered by people with disabilities.

Scribd countered, claiming that ADA governs conduct in public places, but that the web (and cyberspace generally) is not a physical place and thus not subject to ADA requirements. This is manifestly untrue. We may talk about computing as happening in the cloud and in cyberspace, but in truth computing is a terrestrial process that happens on servers in physical locations. Haben Girma, the first Deafblind woman to graduate Harvard Law School, was on the team of public interest lawyers that took on Scribd. After the polite letter (the carrot), her team brought out the stick and went to court. Girma's memoir, *Haben: The Deafblind Woman who Conquered Harvard Law*, offers a powerful account of the 2014 legal battle, which was fought all the way up to the US District Court for the District of Vermont. The stakes were high: even in 2014, tech companies were using a strategy of pretending real-world rules don't apply online. This strategy, designed to protect profits and prevent regulation, enabled widespread discrimination. Girma writes: "Inaccessible websites and apps accelerate the information famine. People who have visual disabilities, dyslexia, and other print-reading disabilities face economic hardships spurred by the lack of access to job applications, health notices, government forms, and educational materials. Technology has the potential to remove barriers, but developers keep designing inaccessible digital services."[5] The precedent was a 2011 case in which the National Association of the Deaf successfully sued Netflix because the streaming giant did not provide captions on streaming video. That decision, rendered in the US District Court for the District of Massachusetts, was the first time a federal court affirmed that the ADA applies to internet-based businesses. The next major step forward came when Girma's team and NFB won their battle, securing access to Scribd's online library for the 61 million Americans with disabilities. Since then, lawsuits about accessible technology have proliferated. The legal mandate for digital accessibility is an important stick. Inclusion is the ethical choice, but when people don't make the appropriate choice, legal recourse is now available to ensure compliance.

Frank Elavsky, a disabled and chronically ill designer who researches accessibility in data visualization, says that a lot of people first experience accessibility from a compliance perspective. "It's doom, you have to bring in an auditor, it's negative," he said. "Instead, think of accessibility as a continuation of communicating, of representing data as an experience. It's an exciting space to be in."[6]

Often, people who care a lot about communicating with technology end up in product design, data journalism, human-computer interaction, or usability. There's a clear line from usability to accessibility, and the concept of interaction design ties the two together. Elavsky said: "Everyone switches seamlessly now between typing, tapping, clicking, speaking, having things read to them. It wasn't like that twenty-plus years ago. The idea of interaction has evolved." He also cautions, however, that accessibility needs to be its own focus, not just a side project in usability. "There is a human rights element that is lost when you talk about accessibility as part of usability," he said. The United Nations has worked for decades to foreground a human rights perspective on disability. In addition to viewing people with disabilities as subjects, not objects, the human rights perspective "entails moving away from viewing people with disabilities as problems towards viewing them as holders of rights," according to a UN report. "Importantly, it means locating problems outside the disabled person and addressing the manner in which various economic and social processes accommodate the difference of disability—or not, as the case may be."[7]

Technological accommodations are abundant, but at the same time they have a long way to go. Almost thirty years into the internet revolution, we're okay at making the mainstream web accessible to people who use screen readers. Thanks in part to the wave of lawsuits that sued organizations for ADA noncompliance, many university websites are now largely friendly to screen readers, which turn website text into either audio or digital Braille. If you want to see what it's like to navigate the web as a Blind or visually impaired person, download a screen reader like JAWS and use it to narrate your web surfing. It will be a totally different experience. Testing websites for accessibility is not an easily automated process.

Auto-captioning tools and transcripts have helped to make online video more accessible to Deaf audiences. Image descriptions in movies and TV allow Blind and low-vision audiences better access to movies and digital video. Some news sites now have a feature that reads the article aloud. Voice technology embedded in voice assistants or navigation software has been helpful for a range of abilities. Subtitles or captioning allow Deaf or hard of hearing users to watch video; they also allow hearing people to watch video in crowded places where sound on a device is inappropriate. Subtitles on films are helpful for language learners. Again, it is the curb cut effect at work.

We achieve remarkable innovation when we open up the pool of people who participate. The design principle at work behind a lot of accessible technology is called universal design. The US government's definition of universal design is "the design and composition of an environment so that it can be accessed, understood, and used to the greatest extent possible by all people regardless of their age, size, ability or disability."[8] Lucy Edwards, a TikTok star who often makes videos poking fun at expectations about Blind people, has a video explaining the difference between accessible and universal design. "Accessible design is specifically focused on the needs of people with disabilities," she explains. She holds up a box of paracetamol that is labeled in Braille, which is helpful to her as a Braille reader. However, she points out that only 10 percent of the Blind population can read Braille, meaning the box is not universal design. "Universal design considers the wide spectrum of human abilities," she says. "For example, my iPhone. It aims to exceed minimum standards in order to meet the needs of a greater number of people. The reason I do what I do is to campaign for universal design. It's not commonplace. I live in a very sighted world. Most of the items I pick up daily, I do not know what they are by just picking them up. It's a lot of guesswork." She holds up a canister of shaving gel and one of air freshener, pointing out that the two are easily confused.[9]

Universal design is the heart of what programmers are taught to do, but many people are limited by their own experience. It's not necessarily malicious; it's unconscious bias and the assumption that your own experience is universal. For example, there aren't a lot of people who stutter who are hired by Silicon Valley tech firms developing voice assistants. Thus voice assistants are unusable by most of the approximately 70 million people worldwide who stutter. This is why it is important to have diverse teams creating technology.

A frequent partner to universal design is design thinking, an innovation process that involves five steps: define the problem, empathize, develop ideas, create a prototype, test/implement. Elise Roy, a Deaf human-centered designer, former lawyer, and motivational speaker, explains her design thinking process in a TED talk called "When We Design for Disability, We All Benefit." It focuses on ways that she has used design thinking to create products, like a pair of safety glasses for woodworking that visually alert the user to pitch changes in the tool that indicate a dangerous kickback is about to happen.[10] Useful innovations like the typewriter, text messaging, audiobooks, remote controls, wide rubber grips on kitchen tools, voice assistants, and closed captioning all stem from designs for disability. "When we design for disability first, we often stumble upon solutions that are not only inclusive, but also are often better than when we design for the norm," Roy said. "This excites me, because this means that the energy it takes to accommodate someone with a disability can be leveraged, molded, and played with as a force for creativity and innovation. This moves us from the mindset of trying to change the hearts and the deficiency mindset of tolerance to becoming an alchemist, the type of magician that this world so desperately needs to solve some of its greatest problems."

Universal design and design thinking communities are not always attuned to intersectional identity, unfortunately. A Black feminist critical technology perspective shows the flaws in their agenda. Ruha Benjamin, in her book *Race after Technology*, shows how design thinking ideas are rooted in white supremacy. "Design is a colonizing project . . . to the extent that it is used to describe any and everything," she writes. "It is not simply that design thinking wrongly claims newness, but in doing so it erases the insights and agency of those who are discounted because they are not designers, capitalizing on the demand for novelty across numerous fields of action and coaxing everyone who dons the cloak of design into being seen and heard through the dominant aesthetic of innovation."[11] An alternative is the design justice approach, which urges that design should be led by marginalized communities and that design should not reproduce structural inequalities. The Design Justice Network Principles, which originate from the 2015 Allied Media Principles (AMP) conference, is a living document that arose from a design community of practice. Sasha Costanza-Chock's book *Design Justice* outlines the principles and explores how "universalist design principles and practices erase certain groups of people—specifically, those who are intersectionally disadvantaged or multiply burdened under the matrix of domination (white supremacist heteropatriarchy, ableism, capitalism, and settler colonialism)."[12] Design can be a tool for collective liberation. The disability justice movement has taken leadership on this issue.[13] The #disabilityjustice hashtag is one place to start learning more; other hashtags for learning about disability include #deaftwitter, #blindtwitter, #a11y, #blindtiktok, #disabilityawareness, and #instainclusion.

In recent years, there have been a few advances in making data visualization more accessible. It's still not easy to create data visualizations if you are blind, but the tools for helping Blind and low-vision people understand data visualizations have become better.

The greatest number of improvements have happened in designing for colorblindness. Most data visualization designers now understand that colorblindness affects many people, and they choose a palette in which the colors don't conflict.[14] Red-green colorblindness is the most common type, affecting an estimated 6 percent of the population.[15] Most of the people affected are assigned male at birth, because the gene that controls colorblindness is on the X chromosome; the condition seems to be more common in white men than other groups. (This pattern of prevalence may be connected to the widespread awareness of the condition.) If a graph includes one line in red and another in green, a red-green colorblind person won't be able to tell which line is which without specific labels. Choosing a different palette is an easy accommodation, and colorblindness-friendly palettes are abundant online. If using red and green together is unavoidable, a designer can adjust red and green shades so that one is very dark and one is very light, as many colorblind people can perceive contrast. Datawrapper, a popular program for making data visualizations, has a helpful built-in feature that reminds the designer to be aware of color choices for colorblind users. It's also common to provide data in a table, in addition to a chart, so that the data is more machine-readable. This is primarily useful for small amounts of data. Providing millions of lines of data in a table is less useful for people using screen readers.

Sonification is a promising field. In sonification, a data visualization designer uses different tones to represent different data values. The simplest sonifications are best. Øystein Moseng, a developer at Highcharts who helped create a sonification program called Highcharts Sonification Studio, has found this is consistent across audiences. "Sonification is a supplementary tool, not a primary tool," he said in a podcast. "The more complex the sonification, the more training and context you need to provide."[16]

Léonie Watson, a data designer who is director of TetraLogical and a member of the W3C Advisory Board, who is Blind, added, "Anything beyond the most simple sonification, like rising or falling tones, I find them quite difficult to interpret."[17]

There is still quite a lot of work to be done in making the digital world more accessible, especially when it comes to new technologies. Haben Girma writes in her book: "Relying on the internet as my primary channel to the outside world constantly throws me against barriers. Many web and app developers ignore accessibility guidelines and the ADA. News feeds are full of images without descriptions, videos without captions or transcripts, and recommendations for new apps to help everyone. In my experience, the word 'everyone' means everyone except disabled people."[18]

Many of the existing technologies have flaws. Alt text, short text descriptions accompanying online images, are a useful tool, but many media sites use image captions as alt text, meaning that a screen reader will merely repeat the image caption. The default in the technology are usually inadequate: the *Washington Post*, one of the most tech-forward media companies, updated their code in 2020 and labeled every image with the alt text "Image without a caption." The engineering team updated it, but only after the issue was flagged by a reader.[19] This is obviously annoying, but it also points to the fact that accessible image descriptions are not part of the workflow for any media organization that fails to create specific image descriptions. Image descriptions can't be automated, which means they have to be created manually by the media organization. This shouldn't be a problem, because image captions are already created manually. However, it does have to be recognized as a priority by the entire organization, including the engineers.

Accessibility technology can often be complicated and expensive. "I started using JAWS when I was seven years old. It is a complicated thing to learn," says Caitlin, a Blind student, in an explanatory video for Challenge Solutions, a website that shares life and technology tips for the Blind and visually impaired. "I do consider myself a proficient user at this point; however, it still confuses me and I don't know everything about it," she said. "There are parts of it that I do not understand . . . there are definitely parts of it that are still over my head. I still get confused as to when I should tab and when I should arrow . . . there are some confusing keystrokes that don't make logical sense."[20] Caitlin is currently a college student who makes videos and podcasts to help other people understand accessibility. If she has been using the software for more than ten years and still finds it confusing, the problem is in the software. She argues that JAWS is so complicated that it shouldn't be the first learning tool given to a student. Still, schools are often eager to hand students a Windows laptop with JAWS installed and be done with it.

Caitlin says that VoiceOver on Mac is easier and is her favorite screen reader. She learned everything she needed to know about

VoiceOver in a weekend. She recommends using an iPad with Blue-tooth keyboard and VoiceOver to make most of the things you need to do as a student easy. It's a fluid experience, she says. The gestures are more intuitive. Everything just *works* with VoiceOver, she says. VoiceOver is her favorite screen reader. It does get confused with ads and pop-ups, however—which is a problem, consider-ing that advertising is the primary monetization strategy of most of the internet. "Most Blind people will tell you an iPhone is the best phone to have," says Caitlin. However, Android phones are typically cheaper. Price concerns are important to consider when designing inclusive technology, given that disability status can be both a cause and an effect of poverty. Similarly, Windows comput-ers tend to be less expensive than Apple computers, and Windows computers are more widely used in schools. JAWS, the most popular screen reader, works best on a Windows device.

Caitlin explains that she recommends getting comfortable with as much tech as possible: "As Blind people, the world is not opti-mized for us. The reality is, sometimes it's going to take three dif-ferent screen readers and two different devices to accomplish a task that a sighted student or employee could do with one device and five minutes. That is our reality right now. You need to be proficient with all the things that could give you a leg up in the world. Think of these things like tools in your toolbox; you want to have as many tools in your toolbox as possible." It's worth noting that Caitlin is a white, relatively privileged college student who is extremely profi-cient with technology, owns multiple computers, and has the time to get familiar with all of them, plus the educational support to do so. This is ideal. It's not necessarily the situation of every Blind per-son. Having a lot of up-to-date technology is expensive and requires a great deal of tech support.

Technology is a crucial component of accessibility, but more of what is currently called cutting-edge technology is not neces-sarily the answer. Autonomous cars, delivery robots, and other so-called smart city technologies are often marketed as solutions that will benefit people with disabilities. The first self-driving car commercial featured a Blind person "driving" a car. This narrative is mostly exploitative. Brand-new technology is rarely accessible. Haben Girma writes of an encounter she had with a delivery robot made by a company called Starship that blocked the path that she and her guide dog traveled in Mountain View, California. It came on the heels of an episode in which a woman in a wheelchair was blocked from getting onto the sidewalk from a busy road because a delivery robot was blocking the curb cut. Girma originally wanted to give the Starship app a chance. She writes: "Thinking the no-contact delivery robots could benefit blind people, I tested the app with VoiceOver on the iPhone. The Starship app refused to fly with VoiceOver, crashing my hopes for a no-contact solution. During a pandemic disproportionately extinguishing disabled lives, the last thing we need is cities adopting tech that excludes blind people and endangers pedestrians with mobility disabilities. . . . Cities and tech companies need to plan for accessibility early in the design process, include disabled people in solutions and review the many published accessibility guidelines."[21] It would be good to stop to make sure everything we have now works for the widest range of people, instead of investing in future technologies that are likely to fail.

Some of the other things that make the world more accessible require a change in mindset. When I started writing this chapter, one thing I didn't understand was whether Blind and Deaf should be capitalized. I learned in my research that there is a difference between being deaf as a fact and being Deaf as a cultural identity. There aren't hard and fast rules. I chose to capitalize the terms, and I used the capitalization preferences of each person I quoted. Another choice was to use identity-first or person-first language. Some people prefer person-first language, as in "people who are Blind," because it centers the individual's experience. Others prefer identity-first language, as in "Blind person," because it centers the identity; it often indicates ownership and pride in the identity. It's okay to ask an interview subject what they prefer, and it's good journalistic practice to do so.[22]

Another important strategy is to presume competence. This brings us back to ethical AI. It is ethical practice to make more accessible technology. However, ethics is not a priority in most computer science education courses. Deborah Raji et al. write:

> The current AI ethics education space relies on a form of "exclusionary pedagogy," where ethics is distilled for computational approaches, but there is no deeper epistemological engagement with other ways of knowing that would benefit ethical thinking or an acknowledgement of the limitations of uni-vocal computational thinking. This results in indifference, devaluation, and a lack of mutual support between CS [computer science] and humanistic social science, elevating the myth of technologists as 'ethical unicorns' that can do it all, though their disciplinary tools are ultimately limited.[23]

Computer scientists, mathematicians, and engineers are exceptionally unprepared to grapple with ethical issues based on their education. Only in the past few years has an ethics course been integrated in the average computer science curriculum. Technochauvinism depends on a perception that people in computer science are special, have more skills, and are smarter or more capable than others. This is problematic at all times, but especially when we look at how non-engineering fields incorporate ethical considerations as part of the core curriculum. Social science and humanities programs routinely teach their students to work through the ethical and social implications of academic inquiry—practices that would serve engineers well.

Incorporating ethics into the computer science and engineering curriculum is essential. Many universities are starting to do it, which is a step in the right direction. Disability and accessibility need to be a focus both inside the ethics curriculum and in the mainstream. Without a focus on disability, technologists end up creating what disability advocate and design strategist Liz Jackson calls disability dongles. "A Disability Dongle is a well-intended elegant, yet useless solution to a problem we never knew we had," Jackson writes. "Disability Dongles are most often conceived of and created in design schools and at IDEO."[24] An example of a disability dongle is a stair-climbing wheelchair. "If you actually talk to disabled people they would tell you that it scares them, that it's too expensive, that they can't afford it, that they simply don't want it," says Jackson. "What they want is access. They want ramps. They want elevators."[25] One starting point for listening to disabled people firsthand is the Disability Visibility Project, a website, anthology, and podcast run by Alice Wong, a disabled activist, writer, media maker, and consultant based in San Francisco. Wong is also the author of a memoir, *Year of the Tiger: An Activist's Life*. In the Disability Visibility Project, Wong collects oral history interviews centered on the lived experience of disability.

Another human strategy is to avoid "inspiration porn." The term, coined by activist Stella Young, refers to images of people with disabilities that suggest that simply living life is inspirational. These are the kind of viral social media posts that show kids with cochlear implants experiencing sound for the first time, or posts with sayings like "The only disability in life is a bad attitude." Disability "is not a bad thing, and it doesn't make you exceptional," Young says. "I use the term porn deliberately, because [inspiration porn images] objectify one group of people for the benefit of another group of people. So in this case, we're objectifying disabled people for the benefit of nondisabled people. The purpose of these images is to inspire you, to motivate you, so that we can look at them and think, 'Well, however bad my life is, it could be worse. I could be that person.'"[26] Instead, a better model is to center stories of people with disabilities, normalize the experience of disability, and call out situations that are *truly* exceptional. "I want to live in a world where we don't have such low expectations of disabled people that we are congratulated for getting out of bed and remembering our own names in the morning," said Young. "I want to live in a world where we value *genuine* achievement for disabled people. . . . Disability doesn't make you exceptional, but questioning what you think you know about it does."

Forward strides in disability justice need to be coupled with more representation and inclusion of BIPOC disabled people. Imani Barbarin, who writes from the perspective of a Black woman with cerebral palsy, notes on her blog Crutches and Spice: "During my time as a disability advocate, I have witnessed disabled people of color, myself included, elevating the need for more representation of diverse disabled people, only to be met with 'well, any representation is good for all of us' and told we need to wait our 'turn.' I have also given talks around the country on disability and race and watched as white disabled people left the room as I discussed these issues."[27]

Whose voices are elevated in the disabled community is also mediated by the digital platforms that determine whose voices are heard in online spaces. When I first started researching this topic, I sought out social media posts associated with a number of disability-related hashtags and looked at the most popular accounts on Twitter, YouTube, and TikTok as well as researching the terms on different search engines. Using online tools alone, I had a hard time finding BIPOC creators or activists. The search algorithms suggested exclusively white creators; the news stories at the top of the search results interviewed mostly white sources. Through the NYU Center for Disability Studies and the intentionally diverse community created by its co-chairs, Faye Ginsburg and Mara Mills, I connected with a wider range of voices. The Center for Disability Studies' report with the AI Now Institute and Microsoft, "Disability, Bias, and AI," illuminated the way that certain voices are privileged and others silenced online.[28] It's not a glitch that BIPOC voices are marginalized when algorithms mediate online discourse; it's a feature of real-life power structures replicating online. In *Algorithms of Oppression,* Safiya Noble shows how search engines reinforce racism by elevating white voices over BIPOC voices. What I saw in my first level of searching was the intersection of ableism and racism. As Barbarin writes: "While the conversation has, over time, shifted to elevating more Black, Indigenous and other people of color, white supremacy is still a major problem for our community. And, while we, as a collective, would like to believe such discussion is behind us, we cannot ignore that racism is deeply rooted in the disability community and that we are currently contending with its effects as representational politics become a self-inflicted wound."[29]

Healing from such wounds is a process. Sometimes the process requires leaving a situation where white supremacy or ableism (or the intersection of factors) is ascendant. Richard Dahan, the #AppleToo activist, has left Apple and has a new job now, working in human resources for the government. "The people at my new job tell me I'm doing great," he said. "I'm surprised—at Apple, I never felt appreciated." He now has the supports and accommodations he needs to thrive in his job.

Dahan wants the concerns of people with disabilities to be part of the emerging conversation about tech and equity. One place this can start is in the language used to refer to the Deaf and hard of hearing community. He says: "We don't like the term hearing-impaired, it implies we are broken. We don't see ourselves as broken. Just because you are disabled, doesn't mean you are broken. It's only a disability when you are oppressed, when you are in a workplace where you are the only Deaf person or you are being oppressed.

"I had this grand picture of Apple. They said they wanted feedback and wanted to grow—but they didn't really want to change and adapt, they didn't care about my needs," he said. "The key is self-advocacy," said Dahan. "A lot of Deaf people are told no when they ask for a sign language interpreter. A lot of time, they did not grow up with the experience I had, of living in a Deaf family and immersed in Deaf culture—they are afraid to stand up. They accept it and move on. It's your right to self-advocate."

7
Gender Rights and Databases

Jonathan Ferguson, then a forty-year-old technical writer at the Ministry of Supply in London, made UK headlines in 1958 when he formally announced his gender transition. "[H]is birth registration has been amended from 'female' to 'male' and his new Christian name inserted into the register," reported the UK's *Daily Telegraph and Morning Post*. This quote, which I read in a paper by scholar Mar Hicks called "Hacking the Cis-Tem: Transgender Citizens and the Early Digital State," has stuck with me because it suggests ease.[1] Someone took a pen and amended a line in the Official Register. In my imagination, it was a fountain pen and written with a flourish, and in that moment Ferguson felt truly seen after years of hiding his true identity. I'm embellishing, but I want it to have been simple and meaningful. These bureaucratic moments, like signing a marriage license or signing the lease on a first apartment, are markers of a life stage transition. I know that it wasn't easy for Ferguson, but I do want it to be easy for the people I love who are going through the same transition seventy years later, when we have achieved significant social progress on some fronts but still have progress to make on others.

In 1958, Ferguson's interaction with the state around his transition involved changing a form and issuing a benefits card. It's not dissimilar to the way that a name or gender change is handled today—though today, the paperwork and expense are far more substantial (and fountain pens are few). From a sociotechnical perspective, it's interesting how the simple pieces of information recorded on a form like Ferguson's benefits card transformed into the design of large bureaucratic computer systems that still govern and control our lives today. The decisions made about how to represent gender in code were efforts (sometimes deliberate) to enforce 1950s ideas about gender on society. Weirdly, we're still living with those retrograde ideas. Despite advances in the understanding of gender and advances in LGBTQIA+ rights, most computer systems still encode gender as a binary value that can't be changed. The next frontier in gender rights is inside databases.

Commercial computing as we know it today started in 1951, when the Census Bureau started running the first commercially produced digital computer, UNIVAC. Back then, gender was generally considered fixed. If you filled out a paper form, it asked for your name and offered you two choices for gender: male or female. You could pick one. Computer programmers used those same paper forms to design computer databases, and when a record was designed for a database, it looked something like this:

```
Firstname
Lastname
Gender (M/F)
Address 1
Address 2
Zip
```

This was how computer database design was taught through the 1990s, when I learned programming. "Back then, nobody imagined that gender would need to be an editable field," one college friend said about how we were taught. Today, we know better. Now, we have a more comprehensive understanding of gender, and an increasing number of companies are embracing inclusive design principles that allow users to self-identify in databases as nonbinary, transgender, genderqueer, and other terms that reflect today's understanding of gender as a spectrum. However, there are artifacts and idiosyncrasies inside computational systems that serve as barriers to implementing truly inclusive design. Most of these problems come from the way that 1950s US and UK social perspectives informed how computer schemas were created. Most people think the issue is changing social norms. It is, but it's also about the way the gender binary is encoded in systems, and how willing engineers are to change those systems.

Many computer scientists and engineers are both personally and professionally committed to the gender binary and cisgender norms, whether or not they realize it. The assumptions coders make without thinking about them can be ones that haunt people for decades. Over and over again, we have seen technical problems arise because the people designing computer systems were committed (consciously or unconsciously) to replicating a rigid, retrograde status quo inherent in the classification systems used by computational infrastructures.[2] "While the gender binary is one of the most widespread classification systems in the world today, it is no less constructed than the Facebook advertising platform or, say, the Golden Gate Bridge," write Catherine D'Ignazio and Lauren F. Klein in *Data Feminism*. "The Golden Gate Bridge is a physical structure; Facebook ads are a virtual structure; and the gender binary is a conceptual one. But all these structures were created by people: people living in a particular place, at a particular time, and who were influenced—as we all are—by the world around them."[3] When same-sex marriage was legalized in the United States, it prompted a database redesign nicknamed Y2GAY. Most databases were set up to allow marriages only between men and women; changing the law required changing those databases to comply. The name Y2GAY is a reference to the Y2K problem, where in the 1990s people realized that most databases and code stored the date as two digits, using an implied prefix of 19. Changing over to the year 2000 was going to screw up an awful lot of code. It was a mess.

All of the popular tech platforms grapple with issues of gender in code. Facebook, which began as a kind of "Hot or Not?" for male undergraduates to rate women, requires users to commit to a gender identity at signup. Facebook famously was among the first social media companies to allow users to change their names and gender identity. Although its software seems to allow users to self-identify as a different gender, the way the system actually stores the data is that each user is recorded (and sold to advertisers) as male, female, or null.[4] The software literally nullifies any gender beyond the binary.

Facebook's parent company, Meta, is not the only Big Tech firm that claims to support the LGBTQIA+ community while falling short on gender inclusivity. For example: face tagging in Google Photos doesn't work well for trans people. If you are trans, photos from before and after your transition may be identified as different people, or the software will ask if multiple photos of you are of the same person. There's no good way to manage being sneak-attacked by photos of your pre-transition self, suggested by the software on your phone. Cara Esten Hurtle wrote of this problem, "The world is full of traps like this for me, whether it's the bouncer who looks at my driver's license and demands a second ID before letting me into the bar, or the unchangeable email address that uses an old name. Trans people are constantly having to reckon with the fact that the world has no clear idea of who we are; either we're the same as we used to be, and thus are called the wrong name or gender at every turn, or we're different, a stranger to our friends and a threat to airport security. There's no way to win."[5]

The reason for this has to do with both hegemonic cis-normativity and math. When you write the kind of computer programs that slot people into neat categories in order to do data analysis, there is a tension between people's messy, shifting identities in the "real" world that rubs up against the sleek empiricism required to do the math that is under the hood in computers. This is most obvious when it comes to the gender binary and binary representation in computer systems.

You know the gender binary: the idea that there are two genders, male or female. (To be clear: I'm describing the concept of the gender binary, not how gender actually works.) Binary code is also the system that powers computers. In a binary numeral system, there are only two numbers: 0 and 1. The numbers 0–4 look like this in binary:

0: 0
1: 1
2: 10
3: 11
4: 100

Computers are powered by electricity, and the way they work is that there is a transistor, a kind of gate, through which electricity flows. If the gate is closed, electricity flows through, and that is represented by a 1. If the gate is open, there is no electricity, and that is represented by a 0. I'm simplifying it dramatically, which will enrage a certain kind of nerd, but here's the gist: this unit of information, a 1 or 0, is called a bit. There are commonly eight bits in a byte and a million bytes in a megabyte (MB). This is how we talk about memory space in computing. Programmers are always thinking about how much memory space a program takes up because space on a computer is finite. When you install a new program, it often asks you to confirm that you recognize the program will take up 3.2 MB of space (or whatever) on your computer. If you don't have enough space available, the program won't be able to install or run.

Different arrangements of bits can be mapped to letters or numbers. In the United States, the most common mapping is called ASCII. In ASCII, the letter A is represented as 01000001. My first name, Meredith, looks like this in the ASCII version of binary representation:

```
01001101 01100101 01110010 01100101 01100100
      01101001 01110100 01101000
```

When a computer stores information about the world, we call that information "data." Data is stored inside a database. In a database, every piece of data has a type, and usually the rules for that type are very strict. In the very simplest form, we can think of data as being of three types: letters, numbers, or binary (0 or 1) values. A binary value is often referred to as Boolean, named after the nineteenth-century mathematician and philosopher George Boole, who invented a system of logic that uses only 1s and 0s. If you want to use data in a computer program, you feed that data to a thing in the program called a variable. Variables also have types, and those types are strictly governed by the rules of a specific programming language. Variable types are slightly different in Python, a contemporary programming language, than they are in C, which is a programming language developed in the 1970s. Unlike human languages, programming languages have very strict grammar and vocabulary. All programming languages have the same essential forms, meaning that they all on some level translate keyboard strokes, mouse movements, variables, data, and other inputs into binary.

Despite all the magical thinking about what computers do, ultimately a computer is a machine built from mined materials, and it uses electricity for calculation. Or, as futurist Ingrid Burrington tweeted, "[I] would like to remind you once again that computers are made out of rocks covered in acid and poison and are deliberately designed to be hard to recycle."[6]

So, in order to store data in our electrically powered poison rocks, we have to declare variables of a certain type inside a database. Speaking loosely, the types are string (meaning text, as in a string of letters), number, or binary (aka Boolean). Boolean variables are used when a value is true or false, and it's represented as 1 or 0: 1 is true, 0 is false. Our database table declaration would look something like this:

```
Firstname  [string]
Lastname   [string]
Gender     [Boolean]
Address 1  [string]
Address 2  [string]
Zip        [number]
```

We'd have to make certain decisions about each field. What type of data goes into each field? How large does each field need to be to hold the intended data? Who can enter the data? Who can change the data? Under what circumstances can the data be changed? Which fields can be edited, and which are fixed? These are all mundane considerations that happen every single time. Usually someone makes a recommendation, then there are a series of lengthy meetings in which people go over the data fields and talk about scenarios like, What happens when someone gets married or divorced, and changes their name? We superimpose human social values onto a mathematical system. The question becomes: Whose values are encoded in the system?

Most of the intellectual history and the dominant social attitudes in the field of computer science can be found in a single, sprawling database published by the Association for Computing Machinery (ACM), the world's largest educational and scientific computing society. The ACM digital library holds virtually all of the canonical papers from computer science journals and conferences. The earliest mention of gender in the ACM digital library comes in 1958, in "The Role of the Digital Computer in Mechanical Translation of Languages."[7] It has to do with translation and pronoun matching in translation. For the next twenty years, any mention of gender has to do with translation. In other words, even though sweeping social change happened in the 1960s and 1970s, including second-wave feminism and the civil rights movement and gay rights and the widespread recognition that sex is biological and gender is socially constructed, academic computer science (and, for the most part, the computer industry) pointedly ignored the topic of gender except to think about how a computer might accurately translate gendered pronouns from one language to another. (For the record, gendered pronoun translation still remains a poorly solved problem today.) Scholars Nikki Stevens, Anna Lauren Hoffmann, and Sarah Florini note a racial dimension to the discourse, writing: "Some of the most prominent works of the database revolution took up 'whiteness' as a kind of unremarked optimum—that is, as the prototype or ideal around which database optimization efforts were (implicitly or explicitly) organized."[8]

In a database, even something as seemingly small as choosing free text entry versus a dropdown has implications. A word occupies more bits than a number and thus takes up more memory space. Today, it's easy to ignore memory concerns, but until the late 1990s, computer memory was expensive. I was taught to write programs that were as concise as possible, and then to refactor those programs down so they took up as little memory space as possible. There's something satisfying about it, writing code that is small and runs fast. Using a Boolean variable is extremely efficient: 0 or 1 takes up less space than 01001101 (M) or 01000110 (F).

If you are designing code for maximum speed and efficiency using a minimum of memory space, you try to give users as few opportunities as possible to screw up the program with bad data entry. A Boolean value for gender, rather than a free text entry field, gives you an incremental gain in efficiency. It also conforms to a normative aesthetic known as "elegant code."

The rigid requirements of "elegant code" enforce cis-normativity, the assumption that everyone's gender identity is the same as the sex they were assigned at birth. It is also a kind of violence against a range of people, including nonbinary, queer, trans, intersex, and gender-nonconforming folks. Being referred to using binary pronouns feels "as though ice is being poured down my back," said nonbinary teen Kayden Reff of Bethesda, Maryland, in court testimony supporting more inclusive gender options in official state documents.[9]

In a paper called "The Misgendering Machines: Trans/HCI Implications of Automatic Gender Recognition," scholar Os Keyes read and analyzed all of the top academic work on automated gender recognition (AGR) from 1995 to 2017 and found that the overall assumption is that gender is binary, immutable, and/or physiological. Keyes writes: "Because AGR treats gender as a binary and physiological phenomenon, there is the potential not just for active harm (misgendering or the enabling of violence) but also erasure; the perpetuation of a normative view that trans people do not exist as a population with needs."[10]

That aesthetic of "elegant code" is specifically exclusionary to someone like Zemí Yukiyú Atabey, who identifies as genderqueer and nonbinary. Atabey's pronouns are ze (where is ze?), zeí (zeí isn't coming today, sorry), and zem (I don't have the tickets, I gave them to zem). "As a nonbinary person, there is no option most of the time," ze says of entering personal information in databases. "There's only male or female, which doesn't fit my reality or identity." Microsoft Word, the program I used to compose this chapter, marked all of Atabey's pronouns with the red squiggly underline, meaning that the people at Microsoft who wrote Word do not recognize Atabey's pronouns as acceptable English words, even though the LGBTQIA+ community has been suggesting the use of ze and hir as pronouns for at least twenty years. The Microsoft engineers chose a dictionary that does not consider ze and hir to be legitimate English words. This is a form of erasure.

I met Atabey when ze was a New York University graduate student. NYU, my employer, is among the most progressive universities when it comes to technological representations of gender identity. Students can change their gender identity in Albert, the student information system (SIS), and professors can see students' preferred pronouns listed on some class rosters.

In the Albert documentation, a distinction is made between legal sex and gender identity. NYU's explanation of it reads: "Legal sex is a person's sex as currently indicated on a birth certificate, passport, or other official document. It may differ from one's gender identity and/or sex assigned at birth. . . . Gender identity is the gender with which a person identifies (i.e., whether one uses man, woman, or more individualized terminology to describe oneself). One's gender identity may or may not differ from one's legal sex which is assigned at birth based on biological characteristics."[11]

Making this change in the database was a complex matter. We may think of computers as nimble and agile, but in reality, changing legacy systems is complicated and expensive. At a university, the student information system is the core system that everything feeds off of, and most SIS systems were set up in the 1960s and haven't been overhauled. There are dozens, if not hundreds, of other systems and programs that feed data back and forth to the SIS every hour of every day. Remember, variables are of strict types. Let's say that you have an old SIS system where you have a field name Sex, of type Boolean. If you then change the field name to LegalSex of type string, and add another field, GenderIdentity of type string, you break the system because the other programs are looking explicitly for Sex, which is a Boolean. You can't pass a Boolean to a function that expects a string, and vice versa.

It is absolutely possible to update university systems to encompass gender identity, just as it is possible to update financial systems, insurance systems, health systems, government systems, and every other system that relies on legacy design. It's a matter of will and funding. Those resources are not evenly distributed throughout large organizations, even at well-funded entities like NYU. "Even though Albert may honor changes to our names, if you go to the dentist on campus or talk to your RA or order books from the bookstore, each place will call you by a different name," NYU undergraduate Joshua Arrayales told me. He was part of an LGBTQ+ club that advocated for the Albert changes. The group's work is not done yet. "Their systems are all so different that changing one doesn't mean changing another," he said. "It gets even more confusing if you've gone by multiple names over the years." It's not uncommon for trans people to try out different names over time. Nor is it uncommon for non-trans people to go by different names and nicknames over time. The computer systems are not built to correspond to how life actually works.

The Albert database transition of the 2020s is reminiscent of what Mar Hicks writes regarding the 1950s: "The massive bureaucracies of the industrialized West now had to face the issue of transgender people transitioning publicly in a way that required state institutions to accommodate them, and it cut to the core of the heteronormative gendered apparatus on which modern states relied . . . new technologies, far from being neutral, were in fact a battleground in the process of defining and stabilizing 'traditional' or normative concepts of gender."[12] Hicks connects the struggles of Jonathan Ferguson and his benefits card to today, writing: "The struggle for trans rights in the mainframe era forms a type of prehistory of algorithmic bias: a clear example of how systems were designed and programmed to accommodate certain people and to deny the existence of others."

The will to enact change is being achieved through lawsuits and legislation. A third gender option, X, on official state documents became available in the United States in June 2021 thanks to the Biden-Harris administration. The change came in response to years of advocacy by groups like the National Center for Transgender Equality. The administration also abolished the onerous requirement that people provide medical documentation to change the gender marker on their passport and consular forms.

Unfortunately, change is slow, and adoption is spotty. Even after gender rights legislation was passed in individual states, more lawsuits followed to force compliance. In March 2021, a group of New Yorkers sued because the databases that control access to Medicaid, food stamps, and other public assistance didn't include the X option that was part of state law. "Any time I need something as simple as food or to make a doctor's appointment, I basically am forced to misgender myself, to be misgendered. And this takes a toll," co-plaintiff Jaime Mitchell said. Mitchell, who is nonbinary, was able to get their birth certificate with an X but was forced to declare themself male or female in order to get public benefits.[13]

Mitchell's complaint is reminiscent of the administrative nightmare that hundreds of trans Britons went through in the 1950s through the 1970s when they outed themselves to the state in order to access the postwar welfare benefits due to them as citizens. The state was transforming into a technocracy, and the lack of an editable gender field or a coherent policy on transition prevented these citizens from accessing benefits. Hicks writes: "The bias against recognizing their identities as valid likewise prevented the government from seeing their complaints as fully real."[14]

In my own practice of allyship, starting with the historical perspective helps me think through what needs to be changed in the modern world. When Ferguson transitioned, he had to fill out paper forms and an official register had to be updated. This is a good way to start thinking through the forms and processes that can be updated to be more inclusive. Today's electronic forms are easier to update than mass-printed paper forms, so one update is to question whether gender needs to be a category on a form at all. If it is necessary to collect gender, systems should make it optional if possible. If not, it should be easy for users to edit and change this field privately, instead of requiring a user to call and have a potentially uncomfortable conversation with a stranger on a customer service line. Forms can provide a range of gender identity options beyond "male" or "female." Lists of options are easily available online. Gender can be a write-in field, rather than a selection from a list.

After thinking about forms, the next step is redesigning systems that the form data goes into. In addition to driver's license systems, public assistance systems, and passport systems, travel systems need to be updated. The US Travel Safety Administration (TSA), which handles airport safety, is a notorious offender when it comes to harassing trans folk. "Transgender people have complained of profiling and other bad experiences of traveling while trans since TSA's inception and have protested its invasive body scanners since they were first introduced in 2010," Harper Jean Tobin, director of policy at the National Center for Transgender Equality, told ProPublica.[15] TSA's full body scanners are programmed with binary representations of gender; some have a pink and a blue button that the agent is supposed to press when a person walks through. Anyone whose gender presentation doesn't match the image expected in the invasive full body scan is flagged. Trans travelers have been forced to strip, endure invasive pat-downs, out themselves, and have been prevented from getting on their flights.

"My heartbeat speeds up slightly as I near the end of the line, because I know that I'm almost certainly about to experience an embarrassing, uncomfortable, and perhaps humiliating search by a [TSA] officer, after my body is flagged as anomalous by the millimeter wave scanner," Sasha Costanza-Chock writes in *Design Justice* of their experience passing through the Detroit Metro Airport. "I know that this is almost certainly about to happen because of the particular sociotechnical configuration of gender normativity . . . that has been built into the scanner, through the combination of user interface (UI) design, scanning technology, binary-gendered body-shape data constructs, and risk detection algorithms, as well as the socialization, training, and experience of the TSA agents."[16] The $110 million that TSA spent on the full-body scanners in 2010 has resulted in trauma and harassment, trauma that is grounded in outdated binary notions of gender embedded in the computing technology. Trans travelers are not the only ones left out of scanners' programming: Black women's hair, Sikh turbans, prostheses, and headscarves commonly trigger these poorly designed systems as well.

The real challenge ahead is in making our current and future technological systems more inclusive while avoiding the mistakes and biases of the past. Lines of code can change the world, absolutely. In celebrating that fact, we need to also look at the way lines of code make culture incarnate and make social change much harder. Computer systems are not just mathematical. They are sociotechnical, and they need to be extensively updated on a regular basis. Just like humans.

8

Diagnosing Racism

If I were to point to a group of professionals who are constantly updating their mental models and technological frameworks to correspond to new social and scientific knowledge, I would choose doctors. As a layperson, I think about the tangle of race in medicine a lot because my own experience with it doesn't fit neatly into computational categories. I am a Black woman with light skin, and people often don't know what racial or ethnic category to put me in. I've been asked if I am Black, white, Puerto Rican, Egyptian, Israeli—the whole spectrum. For the purposes of medical forms, I usually write that I am multiracial. My mother was white, my father was Black, and I want my doctors to be aware of any genetic or epigenetic factors that might be inherited along either family line. As the cultural conversation about race has evolved over the years, I've noticed my doctors adapting their practices.

Medical forms and their descendants, electronic medical record systems, are rarely set up to deal with the nuances of racial identity—or multiracial identity or gender (as discussed in the previous chapter). I remember that on my first visit to my obstetrician's office, in the early 2000s, I had to fill out a paper intake form. I checked two boxes for race, as I usually do. The receptionist called me back up after I handed in my clipboard and forms.

"You checked two boxes for race," she said accusingly. "You're only supposed to check one."

"But that's what I am," I said. I was confused.

"The computer doesn't let you put in two things for race. It only lets you put in one," she said.

"That's a bad system and they should change it then," I said. "I always put down two categories. Put nothing if you have to." The receptionist glared at me. It was not my finest moment, I admit. I was newly pregnant, a little freaked out, and not in the mood to explain to the obstetrical receptionist that race is a social construct, that the transition from paper medical forms to computerized medical records was being mismanaged, and that my racial identity was more complicated than could be captured by the obviously incompetent designers of her hospital system's electronic medical records system.

A couple of months later, I was at the same doctor's office for a routine appointment and happened to glance at my paper chart. It said I was white. I was shocked. Either the receptionist decided to choose for me, and entered me as white, or she entered nothing, and the computer system defaulted to white. I never found out exactly what happened.

"My chart says I'm white, but I'm not," I said to the doctor. "Does that matter for my care?"

"We'll just fix it," she said. She wrote "& Black" next to where it said "Race=white" on the chart, and promised to update it in the computer system. I always remember this moment when people proclaim the virtues of electronic medical records systems. The paper chart allowed my doctor to enter potentially diagnostically relevant information in a way that the computer system didn't, because the computer system was designed by people who had a too-narrow view of race and too much desire to sort people into categories that are easy for databases to understand.

The treatment protocols for medical conditions also have problematic racial legacies. I had a high-risk pregnancy that was handled beautifully at an excellent hospital. The doctors, who were some of the most skilled in the country, struggled to figure out how to decide my various risk factors throughout the process. Should they consider me a Black woman, which would trigger one protocol? Or should they consider me a white woman, which would trigger a different protocol? I feel lucky that my doctors generally understand the ambiguity of the system, run my numbers against both standards, and decide on something in the middle.

The difference between the social construction of race and the epigenetics of race makes it complicated to sort out my individual risk in medical situations. In the United States, Black women are 40 percent more likely to die from breast cancer than white women, according to a CDC report.[1] Obviously, there are some genetic factors to illness, but in general the higher death rate regarding cancer is about social factors such as differential access to care or medical staff disbelieving Black women's pain. Black people's reports of pain tend to be discounted by doctors, which leads to more disability among Black people.[2] During childbirth, Black women are at greater risk of complications because of medical staff's bias (unconscious and otherwise). Even Serena Williams, one of the wealthiest and most recognizable Black women in the world, faced potentially fatal complications after giving birth to her daughter because the medical staff initially ignored her concerns.[3] Soraya Nadia McDonald puts it well in the anthology *Believe Me* when she writes: "It is not enough for Black women to be believed, not when they face an epidemic of historic contempt for their bodies, their reproductive agency, even their ability to feel pain."[4]

Unease with mainstream medicine runs deep in parts of the Black community. There are solid historical reasons for this. For example: in the Tuskegee syphilis study, Black men with syphilis were deliberately denied life-saving medical treatment so researchers could observe exactly how the disease destroys a person's body. In the Henrietta Lacks case, Lacks's cells were taken without her consent and used in experiments that generated enormous wealth for the biomedical industry without any of the wealth being shared with Lacks's family.[5] Pulse oximeters, devices that measure the amount of oxygen in the blood, don't work as well for people with dark skin. Low oxygen levels indicate that a person is seriously ill, meaning that physicians can easily miss a sign of a major health crisis.[6] The list goes on and on. It's unsurprising that there is hesitancy, skepticism, and distrust of the medical establishment inside the Black community.

Cultural critic Damon Young writes of his own skepticism:

> The lack of trust in our nation's systems and structures is a force field; a bulwark shielding us from the lie of the American dream. And nowhere is this skepticism more justified than with the institution of medicine. I don't trust doctors, nurses, physician assistants, hospitals, emergency rooms, waiting rooms, surgeries, prescriptions, X-rays, MRIs, medical bills, insurance companies or even the food from hospital cafeterias. My awareness of the pronounced racial disparities in our health care system strips me of any confidence I would have otherwise had in it. As critics of a recent Saturday Night Live skit suggesting that Black people are illogically set against getting vaccinated pointed out, the vaccine hesitancy isn't due to some uniquely Black pathology. It's a direct response to centuries of anecdote, experience and data.[7]

Black and Brown communities were disproportionately affected by the COVID-19 crisis, especially at the beginning. After that collective trauma in the early days of the pandemic, hesitancy about the COVID vaccine should not have come as a surprise to the public health community.

"Vaccine hesitancy is a symptom of a society that has not taken care of its most vulnerable people, and in fact has exploited and abused them," says Dr. Uché Blackstock, CEO of Advancing Health Equity, an organization she founded to dismantle bias in healthcare. She told me:

It's really no surprise that we are where we are, and we're still struggling to get Black Americans and other Americans of color vaccinated. This is what harm does, this is what structural violence does. When you look at Black communities, we're more likely to be under uninsured or uninsured, so we don't have the same access to health care. When you don't have that pattern of one being able to access health care, when you don't have a trusted provider and then, when you actually *do* interface with the healthcare system and have these very discriminatory experiences, of course you're not going to trust the institution of medicine or health care to help.

The struggles that individual groups have with the medical community matter because today's high-tech medical diagnostic systems are built on top of existing processes—many of which include risk calculations based on problematic conceptions of race. When technological systems are built to diagnose or treat medical issues, as many companies are trying to do, dangerous and wrong ideas about race can be built into the systems. This trend must be reversed, and compensated for, lest existing inequalities in healthcare get worse. Unless we build AI systems with diverse inputs, and we are intentional about eliminating racism at every step of the process, it's easy to predict who will be harmed by AI in medicine.

Consider skin cancer detection and race as a diagnostic category. One of the often-discussed areas for AI is skin cancer detection. In May 2021, Google announced it was launching a new AI-powered dermatology tool that represents a step toward this diagnostic goal. "Today at I/O, we shared a preview of an AI-powered dermatology assist tool that helps you understand what's going on with issues related to your body's largest organ: your skin, hair and nails," the company announced with great fanfare at its annual developer conference.[8] "Using many of the same techniques that detect diabetic eye disease or lung cancer in CT scans, this tool gets you closer to identifying dermatologic issues—like a rash on your arm that's bugging you—using your phone's camera." The tool, which is only available outside the United States, theoretically uses AI to "diagnose" whether a weird thing on your skin is one of 288 different skin conditions.

The company constructed its press release in such a way that it skirts the language of medicalization just enough to stay within the law. They don't claim that the tool diagnoses skin disease, merely that it provides information. Google writes: "Each year we see almost ten billion Google Searches related to skin, nail and hair issues. Two billion people worldwide suffer from dermatologic issues, but there's a global shortage of specialists. While many people's first step involves going to a Google Search bar, it can be difficult to describe what you're seeing on your skin through words alone."[9] It also suggests that this is a pressing issue that technology can save us from, and that Google has special insights because Google Search is more insightful than anything else. It's not.

Skin issues are extremely common, and it's normal to discuss them, unlike some more intimate medical concerns. "What's this weird thing on my arm?" is exactly the kind of thing you can ask your best friend or your sibling or your parent. Then, they usually say, "Put some calamine/Benadryl/cortisone on it, and if it doesn't go away in a few days, call the doctor." If it's urgent, they say, "Go to the emergency room." It's not rocket science. There may be frantic online searching connected to the anxiety about the weird rash, but the general pattern doesn't change.

"Skin, hair, and nail issues" is a very broad category that also includes beauty concerns. Skincare is a huge category in beauty. The cosmetics industry was valued at about $380 billion in 2019. The writers certainly took advantage of this broad category to choose the largest possible numbers to incorporate into the press release and make the scale of the "problem" seem as big as possible.

The stated goal at the end of the Google skin app press release is to get people to search more. So, this is not a selfless diagnostic tool to help people have better skin hair and nail health—it is a tool to get people to search more, which will bring Google more revenue because it makes money from displaying ads every time someone searches a term. The more you use Google search, the more money Google makes. If Google can convince you that AI knows something you don't, you'll search more, and Google will make more money.

The self-dealing would be base enough on its own, but there is a race problem embedded in the money grab. Google trained its skin AI on 64,837 images sourced from 16,114 cases in two US states. The patients are mostly people with light or medium-toned skin. Only 3.5 percent of the images came from patients with darker skin.[10] The training sets, like most of the medical textbooks, include only light skin. This is an echo of the fact that for many years, there was a perception that people with dark skin don't get skin cancer, except on lighter areas like the palms of their hands or the soles of their feet. This is incorrect, but it persists.

Skin cancer looks different on light and dark skin. Medical personnel complain that without appropriate learning materials, it's hard to learn to diagnose skin cancer on a range of skin colors. The skin cancer AIs are likely to work only on light skin because that's what is in the training data. This is a clear bias. "This kind of cavalier attitude that some in tech have had when it comes to health is not surprising. They're rolling out things without necessarily thinking about the public health implications," Dr. Ade Adamson, a dermatologist at the University of Texas's Dell Medical School, told Vice Motherboard about the Google app.[11] "There should have been some caution here. There should have been a prospective [bias and accuracy] study that they put out for us to look at."

In some ways, it is not surprising that Google put out a skin analysis app that works only for people with light skin. Google is the company that categorized photos of Black men as gorillas, shows images of pornography in response to searches for the term "Black girls," and fails to provide trans users ways to manage pre- and post-transition photos or eliminate their deadnames.[12] The company has consistently made choices about its code and its personnel that demonstrate it is steeped in white supremacy. Mainstream medicine swims in these same waters. Despite years of understanding that race is a social construct, race science is still used as a diagnostic. The way race science is encoded in medical decision-making systems can disadvantage or endanger BIPOC patients.[13] For example, until 2021, the threshold for recommending a patient for a kidney transplant differed between Black and white patients.[14] To qualify for the kidney transplant waiting list, a patient needs a glomerular filtration rate (GFR) score of 20 or below. This particular measurement of kidney function means that the kidneys are functioning at only about 20 percent. The score was calculated by an equation that used serum creatinine level, age, sex, race, and body weight. However, Black patients were given a "GFR correction" score of ≥ 1.2 times the baseline, meaning that non-Hispanic white patients qualified for kidney transplants earlier than Black patients. The National Kidney Foundation (NKF) attributed this antiquated race science to "muscle mass." The equations "include a term for the African American race to account for the fact that African Americans have a higher GFR than Caucasians (and other races included in the CKD-EPI datasets and MDRD Study) at the same level of serum creatinine," NKF wrote in a technical explanation on their site. "This is due to higher average muscle mass and creatinine generation rate in African Americans."

The phrase "muscle mass" is a clue here. Conceptually, the fetishization of Black male muscles is linked to the perceived strength of enslaved Black men. By touting an enslaved Black man's strength on the block at a slave market, white kidnappers could drive up the price they charged as they engaged in their despicable trade during the slavery era. Scientific history is filled with nonsense that seeks to document imagined differences between races. "The practice is rooted in a history of work from the fields of anthropometry and eugenics that set out to scientifically prove blacks are biologically distinct and separate from whites," writes Dr. Vanessa

Grubbs, a kidney specialist at the University of California, San Francisco, in a 2020 issue of the *Clinical Journal of the American Society of Nephrology*, "[w]ork like that of well-known and respected (at the time) physician Samuel Cartwright's 1851 Report on the Disease and Physical Peculiarities of the Negro Race, in which he claimed, 'The darkness of the Negro's skin pervaded his membranes, muscles, tendons, fluids and secretions, including his blood and bile.'"[15]

Patients, doctors, scholars, and activists have been advocating for years to get the standard GFR equations changed so that Black people are not disadvantaged on the transplant list.[16] Their antiracist advocacy led the National Kidney Foundation and the American Society of Nephrology to form a joint task force in 2020 to re-examine the use of race in the standard mathematical formula used to estimate GFR. In September 2021, the task force published a new formula that does not use race as a factor. "Unlike age, sex and body weight, race is a social, not a biological construct," the organizations wrote in a press release. "Including adjustment for race in these eGFR equations ignores the substantial diversity within self-identified Black or African American patients and other racial or ethnic minority groups."[17] This is tremendous progress, and the groups should be commended for reckoning with their racist past. How many months or years will it take for every laboratory and every medical record system and every algorithmic medical system to adopt the new formula? This remains to be seen. Will the change be accompanied by patient and practitioner education about the problematic history of race-based eGFR, plus better access for Black communities to preventive kidney health care, diagnosis, timely nephrology referrals, high-quality dialysis services, and transplant services?[18] This is also an open issue.

Density also shows up in myths about Black women's breasts, which are widely believed to be more dense than other women's breasts. Radiologists use a system called Breast Imaging Reporting and Data System (BI-RADS) to classify breast density into four categories based on a breast's mix of fatty breast tissue, glandular tissue, and fibrous connective tissue. Dense breasts are a risk factor for breast cancer. My friend L, a white woman who was diagnosed with breast cancer in Tennessee a few years ago in her thirties, was repeatedly told that her breasts were extremely dense. It made it hard to see her cancer, the doctors said. They marveled that her breasts were so dense because she wasn't Black. In reality, breast density varies and is genetic and does not correspond to race. People with dense breasts are often made to feel like their breast density is a problem in imaging. It's as if the doctors and technicians blame the breasts for it being hard to read the films. Instead, it would be better to blame the technology, which is painful and doesn't work as well on dense breasts.

In addition to devaluing BIPOC lives, race "correction" in medicine has direct economic effects that uphold white supremacy. "The use of race-based tables to lower damages for BIPOC claimants in tort lawsuits devalues their lives & is directly related to race correction in medicine," said Dorothy Roberts, a University of Pennsylvania law school professor and the author of *Fatal Invention: How Science, Politics, and Big Business Re-create Race in the Twenty-First Century*. She writes:

> It gives added incentive for corporations to locate hazardous facilities near Black communities. Here's the connection: race correction in medicine embeds the assumption that Black people have bodies that naturally & categorically function worse (eg, lower lung or cognitive capacity), so have been harmed less than white people with the same injuries. This is why, based on race-norming, the NFL denied claims by former Black players for dementia caused by concussions while paying white players with the same level of injury. And why companies argued Black workers should get lower damages in asbestos lawsuits. So race correction not only perpetuates racist ideas about Black bodily difference traced back to slavery, disqualifies Black patients from equal care (eg, eGFR/kidney transplants), but also lowers monetary compensation corporations owe for damaging our health.[19]

Race "correction" also shows up in other medical contexts like pulmonary function thresholds for lung transplants and in who is recommended for a vaginal birth after a previous birth via caesarian procedure.

Given this history, it is dangerous to transfer diagnostic methods to algorithmic systems without first scrutinizing whether the diagnostic methods are disadvantaging certain groups. The current boom in AI, and newer techniques like deep learning and neural nets, has fueled the current enthusiasm for computer-assisted diagnostic medicine. The goal of using AI to diagnose goes back almost as far as AI itself.

AI began in 1956, when a group of scholars gathered at the Dartmouth math department to name their new field and create its central questions. At first, the field was focused on problems like how to get computers to beat humans at chess. This group of scholars liked chess and thought that beating other people at chess was a marker of human intelligence. If they could make a computer that could beat humans at chess, they reasoned, the computer would be artificially intelligent. That did not prove to be true. When computer scientists eventually managed to make a machine that could beat the world's best human chess player, it was a great accomplishment but was obvious that the computer was not reasoning; it was just churning through math and data. As the field progressed, there was a lot of rhetoric about how AI was going to replace human experts. The first AI boom, in the 1980s, was focused on expert systems. These systems, which were never perfected, were envisioned as a kind of doc-in-a-box that would take in scans and output diagnoses and treatment plans. The failure of the 1980s AI paradigm led to a few years of de-investment, which people call the AI winter. In the 1990s, advances in neural nets kicked off a new season of popularity for AI, and machine learning methods became all the rage. Deep learning and neural nets are both considered types of machine learning.

Geoffrey Hinton, a University of Toronto computer scientist often called the father of deep learning, thinks that radiologists should just give up because AI is the future. "I think that if you work as a radiologist you are like Wile E. Coyote in the cartoon," Hinton told Siddhartha Mukherjee in a 2017 *New Yorker* interview. "You're already over the edge of the cliff, but you haven't yet looked down. There's no ground underneath." Hinton believes that deep learning, his field of expertise, is the breakthrough tech that will change diagnostic medicine forever. "It's just completely obvious that in five years deep learning is going to do better than radiologists," he said. "It might be ten years. I said this at a hospital. It did not go down too well."[20]

This is an excellent example of a recurrent problem in computer science, specifically among machine learning experts. They think they know it all. "We as computer scientists tend to have a lot of hubris about anything we come across. We think that we can invent it better than anybody else has done in the past, whether that's a new technology or whether that's coming into a field and disrupting it, or whether that's a concept like fairness and bias," said machine learning researcher Jonathan Frankle on a 2021 panel about values in engineering. "We have a tendency to show up and ignore decades or centuries of prior work and expertise from people who have been thinking about these questions very carefully. We redefine or make up our own perspectives on these and only slowly discover the prior work as experts come into the room and perhaps educate us slowly. But even then, I don't think we often listen."[21]

This is exactly what's going on with Hinton. It's five years after his grandiose statement, and deep learning is not better than radiologists. The history of technology is littered with overblown statements like his, statements that are just as problematic. Journalists need to stop printing these kinds of forward-looking claims, and we the public need to stop believing them. The easiest strategy is that we need to start "calling bullshit" on claims about future technology and a rosy tech-enabled future.

Biologist Carl T. Bergstrom and information scientist Jevin West teach a class called "Calling Bullshit" at the University of Washington, and published a book with the same name. "Bullshit involves language, statistical figures, data graphics, and other forms of presentation intended to persuade by impressing and overwhelming a reader or listener, with a blatant disregard for truth and logical coherence," in their definition. They offer a simple, three-step method for bullshit detection, which involves these questions:

1. Who is telling me this?
2. How do they know it?
3. What are they trying to sell me?[22]

Hinton is a deep learning researcher who is trying to sell the journalist on his expertise, so he can enhance his own reputation. He's involved in commercializing deep learning technology in several ways. In addition to his work at the University of Toronto, Hinton also works for Google Brain, which he joined in March 2013 when Google purchased his company, DNNresearch Inc. This is a strategy called an acqui-hire, where a company buys a company not just for the technology but in order to hire the company's staff. The deal was documented in a university press release: "The Google deal will support Prof. Hinton's graduate students housed in the department's machine learning group, while protecting their research autonomy under academic freedom. It will also allow Prof. Hinton himself to divide his time between his university research and his work at Google. . . . Professor Hinton will spend time at Google's Toronto office and several months of the year at Google's headquarters in Mountain View, CA. This announcement comes on the heels of a $600,000 gift Google awarded Professor Hinton's research group to support further work in the area of neural nets."[23]

Believe Hinton on deep learning. Be skeptical of Hinton on cancer detection or other highly specialized real-world applications of deep learning. The boundaries between commerce and academia are porous in machine learning nowadays—perhaps too much so. Selling one's machine learning or data science expertise is the easiest way to get rich as a computer scientist. "Computer science research in its current state is basically a free corporate research lab," said Brown University computer scientist Seny Kamara, who is one of roughly three Black cryptography researchers in US academia.[24] Cryptography, like its parent field of computer science, is not very diverse. In a talk at the Crypto 2020 conference, Kamara points out the ways that cypherpunk, an open-source movement touted as an alternative to Big Tech, fails to prioritize marginalized communities in its definition of who gets liberated by technology. "Cypherpunks are concerned with personal freedom, which of course is important; some of my work is on surveillance so clearly I think this is very valuable," said Kamara. "The focus of that community often is personal freedom with respect to government and intelligence agencies. That's a very libertarian perspective. . . . What I'm concerned about is cryptography (and more broadly, technology) that is concerned with fighting oppression and fighting violence from law enforcement, from the police, from the FBI, from ICE. From social hierarchies and norms, from domestic terrorists, from Nazis and the alt-right and white supremacists and religious fanatics. I don't know if the technology we're producing is concerned about this." Social priorities lag inside computer science; there are economic incentives to ignore social issues; and the pseudoscience of race has confused the scientific discourse. It's a mess.

The same kind of mess could be repeated inside the emerging field of precision medicine, which, loosely speaking, seeks to use data to tailor treatments to patients. Precision medicine has been used successfully in cancer care. Doctors can sequence tumor DNA to ascertain which kind of therapy might work best against a tumor. "In clinical precision medicine, there has also been some success in pharmacogenetics, a field that uses an individual's genetic sequence to tailor dosages of medications," write Kadijah Ferryman and Mikaela Pitcan in a 2018 report from the nonprofit group Data & Society titled "Fairness in Precision Medicine." They warn that data-driven precision medicine is susceptible to bias at multiple points, from bias in datasets used to train precision medicine models as well as bias from outcomes.[25]

It's clear that simply reproducing existing diagnostic processes in the digital world is not the answer. But how could we build tech that is both ethical and accurate and takes advantage of any digital gains while not perpetuating racist or sexist or ableist consequences?

One approach involves looking at three different kinds of bias: physical bias, computational bias, and interpretation bias. This approach was proposed by UCLA engineering professor Achuta Kadambi in a 2021 *Science* paper.[26] Physical bias manifests in the mechanics of the device, as when a pulse oximeter works better on light skin than darker skin. Computational bias might come from the software or the dataset used to develop a diagnostic, as when only light skin is used to train a skin cancer detection algorithm. Interpretation bias might occur when a doctor applies unequal, race-based standards to the output of a test or device, as when doctors give a different GFR threshold to Black patients. "Bias is multidimensional," Kadambi told *Scientific American*. "By understanding where it originates, we can better correct it."[27]

A multidimensional framework such as Kadambi suggests can be useful if it is effectively integrated into everyday practice and implemented along with appropriate training. There are also mathematical ways of addressing bias. I picked this specific article to look at because it's an example of how a scientific paper can be accurate in spots but also wrong. Kadambi spends the first part of his paper correctly citing examples of bias in medicine and looking at the cognitive fallacies that lead to the bias failing patients. Looking at a multidimensional solution is also a good strategy. However, then Kadambi recommends looking at different kinds of bias and deciding what is an acceptable threshold, which is a popular idea among computer scientists but not among the people affected. I disagree with Kadambi on this point. I want to normalize *not* using technology when the technology is impossible to make fair. "Computer systems can have an acceptable level of bias" is an idea, an argument, not a fact—for the moment, at least. It is contentious. It is presented in this paper as a fact, backed up by the writer's expertise. This should not have made it through the editing process.

Kadambi then goes on to say race is real. This is also not a fact. Race is a social construct; it is not innate, not a medical fact of the human body like bones. This is another claim that shouldn't have made it through the editing process. That these two grave mistakes made it into one of the most reputable scientific journals is due to another kind of cognitive bias, having to do with expertise.

One interesting mistake most people make about experts is that we assume their expertise translates to other realms. Kadambi is a computer vision expert. The editor(s) assumed that he knew what he was talking about with race and with bias, because he knows what he is talking about with computer vision. This is a mistake. In fact, considering how little the typical engineering and computer science curriculum deals with bias and ethics, we should arguably discount the bias/ethics opinions of computer scientists and engineers *more* than social scientists when it comes to social issues.

Cognitive biases are everywhere, so much so that every knowledge field generally has a catalogue of how cognitive biases manifest. Some of the useful search terms are "cognitive bias," "implicit bias," "unconscious bias," and, for medicine specifically, "diagnostic error and bias."[28] We all have unconscious biases. Many of us are working every day to overcome them and become better people, but none of us is perfect. Doctors' unconscious and implicit biases unwittingly perpetuate health care disparities. "We still see how much race plays a role in health outcomes," Dr. Blackstock told me. "I always say it's not race, it's racism. It's the environment that's created, that really impacts health."

Diagnostic technology built on layers of scientifically unsound, racialized practices will be bad, and there's no way around it. There's no such thing as a perfectly unbiased person who will be the perfect system designer, just as there's no such thing as perfectly unbiased data. In a lot of these cases, we must ask: Why use inferior technology to replace capable humans when the humans are already doing a good job? In the hopefully rare cases where humans are not doing a good job, are there noninvasive or nonsurveillance-based ways to use technology to help them do better? And if we must use inferior technology, let's make sure to also have a parallel track of expert humans that is accessible to everyone regardless of economic means, and fund or prioritize it so that the humans can subvert the race correction formulas that disadvantage BIPOC communities. "Since the slavery era, race correction and norming have treated the effects of racism on Black people's bodies (or imagined difference) as innate 'peculiarities of the Negro,' thereby reproducing white supremacist ideology, injury & injustice," said Dorothy Roberts, the UPenn professor, on Twitter.[29] "That's why they should be abolished."

9

An AI Told Me I Had Cancer

Do you know what they say about using AI to diagnose tumors? It'll grow on you. There's another AI story I want to get off my chest. In late 2019, I went in for what I thought was a routine mammogram. The radiologist reading my images told me there was an area of concern and that I should schedule a diagnostic ultrasound. At the ultrasound appointment a few days later, the tech lingered on an area of my left breast, and frowned at the screen. I knew then it would be bad. Another mammogram and several doctor visits later, it was certain: I had breast cancer.

I was scared. On top of being afraid of the cancer, I was afraid of being a Black woman with breast cancer. As I mentioned earlier, breast cancer death rates are 40 percent higher among Black women than white women in the United States. My personal history wasn't reassuring, either. I was the same age my mother was when she was diagnosed with breast cancer. She had died, painfully, five years after her diagnosis. My mother was white, healthy, with good medical insurance, and had a kind of cancer that is curable in 90 percent of cases. She died anyway. I was looking at a different kind of cancer, and I am Black. Was I 40 percent more likely to die from my cancer than my late mother? I was extremely worried.

Everybody freaks out after a cancer diagnosis. The specific *way* that you freak out tends to be consistent with your personality. This explains why I spent the first days after my diagnosis in shock. Then, I read everything on the internet about breast cancer, then called all my friends and family, and then read every forum post on breastcancer.org for the past three to five years, and did an extensive search of relevant research papers in the biomedical literature database, then created a spreadsheet of things I needed to buy for the recovery period after my mastectomy: a reclining chair, a foam wedge for bed because I wouldn't be able to lie flat for a while, and a small apron, the kind that restaurant servers wear, to hold the drains that would be in my chest for a week after the surgery. Over-functioning is my primary coping mechanism.

Because I like to know everything that is happening to me medically, and because I think the poor user interface design of electronic medical record systems leads to communication problems among medical professionals, I always poke around in my online medical chart. Attached to my mammography report from the hospital was a strange note: "This film was read by Dr. Soandso, and also by an AI." An AI was reading my films? I hadn't agreed to that. What was its diagnosis?

I had an appointment the next day for a second opinion, and I figured I'd ask what the AI found. "Why did an AI read my films?" I asked the surgeon the next day.

"What a waste of time," said the surgeon. They actually snorted, thinking the idea was so absurd. "Your cancer is visible to the naked eye. There is no point in having an AI read it." They waved at the computer screen, which showed the inside of my breast. The black and white image showed a semicircle on a black background filled with spidery ducts, with a bright white barbell marking the spot of my diagnostic biopsy. The cancerous area looked like a bunch of blobs to me. I was grateful in that moment that this doctor was so expert and so eagle-eyed that they could spot a deadly growth in a sea of blobs. This was why I was going to a trained professional. I immediately decided this was the surgeon for me, and I signed the form agreeing to an eight-hour operation.

The day of the surgery came. I went to the hospital at the crack of dawn. In the taxi, the driver was playing a news station on the radio, and there was a bulletin about a virus in Wuhan, China. He coughed. We looked at each other uncomfortably. I got to the hospital and got prepped. The staff was helpful and kind. In the operating room, I looked up at the starburst of lights and counted backward from ten.

They gave me morphine for the pain afterward. I don't remember much between the surgery and the next day. That day was agonizing; I couldn't stand up straight, and shuffling across the room was pure torture. The pathologist reported that the lymph nodes they removed from my armpit were clear, meaning the cancer had not hitched a ride on my lymphatic system to spread elsewhere in my body. I had a new torso and a fake left breast, and after a few days in the hospital they sent me home to recover. I felt fortunate to have the kind of healthcare that pays for me to be treated by my choice of cancer specialists, and I was keenly aware of the economic and social privilege that had paved my way through the cancer-industrial complex. I was grateful.

As soon as the morphine wore off (and the other painkillers), New York City shut down for the first wave of the COVID-19 pandemic. I was weak, I was mildly stoned on Valium (a common treatment for muscle tremors after a mastectomy), and I was curious about the breast cancer AI. In between episodes of forgettable television shows, stuck in my new reclining chair, I started reading. I couldn't walk very well and there was a pandemic devastating my city. Reading about a *known* deadly disease felt more reassuring than thinking about COVID-19, which was then an *unknown* deadly disease.

Fast-forward a few months, and I was mercifully cancer-free and mostly recovered. I got a clean bill of health one year out, I was vaccinated against COVID-19, and I decided to investigate what was really going on with breast cancer AI detection.

I had found out about the cancer detection AI because I was nosy and read the fine print. Patients today often don't know that AI systems are involved in their care. This brings up the issue of full consent. Few people read the medical consent agreements that we are required to sign before treatment, much like few people read the terms and services agreements required to set up an account on a website. Not everyone is going to be thrilled that their data is being used behind the scenes to train AI or that algorithms instead of humans are guiding medical care decisions. "I think that patients will find out that we are using these approaches," said Justin Sanders, a palliative care physician at Dana-Farber Cancer Institute and Brigham and Women's Hospital in Boston. "It has the potential to become an unnecessary distraction and undermine trust in what we're trying to do in ways that are probably avoidable."[1]

I wondered if an AI would agree with my doctor. My doctor had saved me from an early grave; would an AI also detect my cancer? I devised an experiment: I would take the code from one of the many open-source breast cancer detection AIs, run my own scans through it, and see if it detected my cancer. In scientific terms, this is what's known as a replication study, where a scientist replicates another scientist's work to validate that the results hold up. If the experiment can't be reproduced, the scientific conclusions are called into question. I assumed that the software would take into account my entire electronic medical record, and I planned to also test to see if the software was sensitive to race, which is often used as a medical diagnostic category. Would changing the race listed in my chart change the diagnosis?

My personal replication study started with understanding the diagnostic process. Then I had to look at the code and the training data used to train the AI models. AI cancer diagnosis, like regular diagnosis, starts with the kind of body part and the kind of scan.

Each body part has a customized medical imaging process. Some of the well-known medical imaging techniques for a breast are mammography, MRI, and ultrasound. Ultrasound is the same imaging process used to look at an unborn baby. The technician uses a machine that bounces sound waves off internal tissues, and the sound waves are turned into an image. MRI, or magnetic resonance imaging, uses magnets and radio waves to create images that form a cross-section of images of a person's innards. Mammography is a highly specialized X-ray, like what you'd get if you broke your arm, except they compress your breast between two plates in order to get a comprehensive image. There is a new kind of 3-D mammography called digital breast tomosynthesis (DBT), which takes multiple pictures of each breast from multiple angles using an X-ray tube that moves in a half-circle around the breast. The images from the seven-second exam are assembled by the computer into a 3-D image that a medical professional can then read. DBT uses less compression than traditional mammography, which in the best-case scenario will lead to more people getting diagnostic mammography. Many people avoid routine mammography because it is painful to have the technician, who is usually quite nice but a total stranger, lift your naked breast onto a plate in the machine and squeeze it flat as a pancake.

In breast cancer, as in some other kinds of diagnosis, there are internal and external diagnostic procedures. Mammography and MRI and ultrasound evaluate what's happening internally. External analysis is achieved by digital manipulation. Digital doesn't mean using a computer, but digital as in *with digits*, meaning fingers. If you have a lump in your breast, the doctor feels it with their fingers and assesses the texture of the lump. Benign cysts feel smooth and typically have regular shapes. Malignant masses normally feel lumpy and irregular. Regular breast self-exam is recommended so that people can detect masses and get them checked out as necessary.

Traditional diagnostic results are the foundation for AI diagnostic systems. AI diagnostics is a fast-growing sector because there is a lot of enthusiasm about potentially using AI in the future. Sometimes this takes the form of claiming to make diagnosis more accurate. Sometimes people are open about their goal of replacing doctors and medical personnel, usually as a cost-cutting measure. The way you figure out what is going on in state-of-the-art computational science is by looking at open-source science. All of the people developing proprietary AI methods look at what's happening in open science, and most use it for inspiration. Microsoft's GitHub, the most popular code-sharing website, hosts most of the available code. ArXiv.org hosts most of the relevant papers, whether in preprint or in finished version. Kaggle, a popular site for data science competitions, is another place to look. GitHub offered 1,350 different public projects on breast cancer detection when I searched in March 2021.

When you read enough of the papers and code, some common themes emerge in both methods and sources. Kaggle has almost 2,000 notebooks and twenty-seven competitions using the University of California, Irvine's Machine Learning Breast Cancer Wisconsin (Diagnostic) Data Set. The UCI repository is quite popular. It includes a dizzying variety of datasets, like 13,611 images of seven different kinds of dried beans used to build a computer vision system for uniform seed characterization. The Wisconsin breast cancer dataset was compiled from real patient images gathered from patients in a Wisconsin hospital system. Surgeons and computer scientists collaborated to compile and label the images, which they assembled into datasets and donated to UCI for research purposes. We tend to think of machine learning as a field where people use a wide variety of data, but in reality, most people are drinking from the same well because the number of available big datasets is not large.

All AI cancer diagnosis starts with a dataset. The Wisconsin dataset I looked at features an ID number and a diagnosis: M for malignant and B for benign. Additionally, ten features are computed from a digitized image of a fine needle aspirate of a breast mass. They describe characteristics of the cell nuclei present in the image. The features are numerical and include mathematical values for radius, texture, perimeter, area, concavity, symmetry, and fractal dimension.

I realized that I had a colleague in the data science department who was building a breast cancer detection AI. He also lived downstairs from me in my apartment building. Had his AI read my films because the hospital had included my mammography images in a clinical dataset used for research? Had he personally seen my films? How did I feel about my neighbor seeing images of my breast—even if it was just the inside? I had consented to my films being seen by the people involved in my care, but I did not consent to my neighbor checking out my boob.

I decided to lean into the weirdness and run my own medical images through my neighbor's breast cancer detection code to investigate exactly what the AI would diagnose. I wanted to use neural nets because they seemed more accurate. My neighbor's AI used neural nets and had good laboratory performance.

The experiment seemed straightforward: I planned to download my scans, download the detection code, run the scans through the code, find out if the program detected cancer, and interview my neighbor about the results in order to explore the state of the art in AI-based cancer detection. In more scientific terms, I laid out a replication study with a sample size of one.

The plan went off the rails immediately.

I saw my scans in my electronic medical record (EMR). I tried to download them. I got an error. I tried to download with the data anonymized, per the options. The EMR offered me a download labeled with someone else's name. I couldn't check if the images were mine or this other person's, because the download package didn't have the necessary files to open the package on a Mac, which was my primary computer. Medical images are standardized in a format called DICOM, which theoretically allows the files to be read on multiple platforms and allows patient information to be attached to the files. DICOM files are usually optimized for PCs, in my experience. I clicked some different options and I eventually got images with my own name, but the package never opened on a Mac.

I tried to open the package on my son's gaming computer, a PC. I got the error message that some information was missing, and the reader packaged with the files wouldn't launch. After a few days, I concluded that the download code was broken. I called the hospital, which put me in touch with tech support for the portal system. I got on the phone with tech support, escalated to the highest level, and nobody was interested in fixing the code or investigating. They offered to send me a CD of the images. "I don't have a CD drive," I told the friendly person in tech support. "Nobody has a CD drive anymore. How is there not an effective download method?"

"Doctors' offices keep CD drives for reading images," she told me. I was nearly incandescent with frustration by this point.

Back in my office, I resorted to the most low-tech strategy I could imagine. On my Mac, I took a screenshot of the images in my EMR. I was unimpressed with the EMR technology. I sent the images to my research assistant, Isaac Robinson, who downloaded the detection code from my neighbor's repository on GitHub, the code-sharing website. After a few days of fiddling, Robinson got the code going.

I should say that I picked my neighbor's code because he is a smart guy and I figured his code would be state-of-the-art. He's a well-respected researcher in the field and I respect him as a colleague. I could have picked any other open-source cancer detection code. My neighbor's name is Krzysztof Geras, and the code accompanied his 2018 paper "High-Resolution Breast Cancer Screening with Multi-View Deep Convolutional Neural Networks."[2]

I had assumed that the software would look at my entire medical record and evaluate whether I had cancer, like a doctor looks at a patient's whole chart. Wrong. Each cancer detection program works a little differently and uses a different specific set of variables. Geras's program takes two different views of a breast. They are semi-circular images with light-colored blobs inside. "It looks like mucus," Robinson said after looking at dozens of these images in order to set up the software.

Breasts vary widely, but mammography is largely standardized. Mammography involves a set of images taken at different angles. Each angle has a name. Roughly speaking, there are two mammography techniques: standard views and supplementary views. Mediolateral oblique (MLO) and craniocaudal (CC) views are standard, and are taken at nearly every mammography. Supplementary views are extras, and they are taken if the tech or radiologist sees an area of concern or wants to examine a particular area more closely. Mammography is constrained by the physics of the human body. The radiological picture is generated by shooting radiation through a tube at a plate. If you put a body part between the plate and the tube, the plate will show an image of the body part. The easiest way to see what's happening inside a breast is to flatten it and see what happens to the internal structures. The breast gets flattened between two plates. Photos are taken at different angles. You get different images if the plates are vertical, versus horizontal, versus at different angles to the body. Any time there is standardization in medical tests, it is a target for automation or AI.

The way an AI generates a prediction is very different from the way a doctor generates a diagnosis. A doctor looks at evidence and draws a conclusion. A computer can generate a prediction that a given situation is malignant. A prediction is different from a diagnosis. Humans use a series of standard tests to generate a diagnosis, and AI is built on top of this diagnostic process. Some of these tests are self-exam, mammography, ultrasound, needle biopsy, genetic testing, or surgical biopsy. Then, you have options for cancer treatments: surgery, radiation, chemotherapy, maintenance drugs. Everyone gets some kind of combination of tests and treatments.

I got mammography, ultrasound, needle biopsy, genetic testing, and surgery. My friend, diagnosed around the same time, detected a mass in a self-exam. She got mammography, ultrasound, needle biopsy, genetic testing, surgical biopsy, chemotherapy, surgery, radiation, a second round of chemo, and maintenance drugs. The treatment depends on the kind of cancer, where it is, and what stage it is: 0–4. The tests, treatment, and drugs we have today at US hospitals are the best they have ever been in the history of the world. Thankfully, a cancer diagnosis no longer has to be a death sentence.

Fine needle aspiration, also called a needle biopsy, is often the first test you get after a mass is detected by mammography, ultrasound, or manual exam. It's pretty simple: a doctor sticks a long needle into the mass to get a sample, then runs some tests on the cells. One of the ways they do the test is kind of funny. There is a table with a hole in it, and you lay down on your stomach and stick your breasts through the hole. The doctor pushes a button to send the table up in the air about to the height of the top bunk of a bunk bed. Then, the doctor scoots their chair underneath the table and sticks a long needle up into your breast. The UCI Wisconsin Breast Cancer Diagnostic Dataset, which I mentioned earlier, contains 569 results of fine needle aspiration of breast masses. In the Wisconsin dataset, numerical features are computed from a digitized image of hundreds of needle biopsies.

Other algorithms are trained on other datasets. Geras's code that I downloaded was based on the NYU Breast Cancer Dataset, which includes screening mammograms for patients aged nineteen to ninety-nine.[3] Specifically, the dataset had 1,001,093 images sourced from 229,426 exams of 141,473 patients at NYU Langone Health imaging facilities over seven years. This is big for a medical dataset, and at the time of writing was the largest breast cancer screening dataset available. My own scans were not in the dataset; I was relieved to see that my fears of boob photo distribution were overblown. The dataset was created well before my diagnosis, and explicit consent was obtained from all participants. Medical data, because it is protected by HIPAA, is harder for researchers to obtain than, say, social media data (which can be easily scraped from the web or purchased from data brokers). The patients' self-reported race and ethnicity in the dataset were 50 percent Caucasian, 30 percent African American, 15 percent Hispanic, and 5 percent Asian. This reflected the population of the NYC metropolitan area, where the images were taken. It took four weeks to train the model, using state-of-the-art hardware. I don't want to think about how much it cost to run that much analysis; it must have been hideously expensive and energy-intensive, incurring the dramatic environmental costs of all AI.[4]

Because Geras and his collaborators pre-trained the model and put it online, all Robinson and I had to do was connect our code to the pre-trained model and run my scans through it. We teed it up, and . . . nothing. No significant cancer results, nada. Which was strange because I knew there was breast cancer. The doctors had just cut off my entire breast so the cancer wouldn't kill me.

We investigated. We found a clue in the paper, where the authors write, "We have shown experimentally that it is essential to keep the images at high-resolution." I realized my image, a screenshot of my mammogram, was low-resolution. A high-resolution image was called for.

Robinson discovered an additional problem hidden deep in the image file. My screenshot image appeared black and white to us, like all X-ray images. However, the computer had represented the screenshot as a full color image, also known as an RGB image. Each pixel in a color image has three values: red, green, and blue. Mixing together the values gets you a color, just as with paint. If you make a pixel with 100 units of blue and 100 units of red, you'll get a purple pixel. The purple pixel's value might look like this: R:100, G:0, B:100. A color digital photo is actually a grid of pixels, each with an RGB color value. When you put all the pixels next to each other, the human brain forms the collection of pixels into an image.

In art, the most famous example of this is Georges Seurat's pointillist painting *A Sunday Afternoon on the Island of La Grande Jatte*. From a distance, the painting shows people at leisure on a riverbank, notably, a woman in a bustle holding a parasol. Up close, you can see that the painting is composed using tiny dots of paint. Every digital image is like a pointillist painting in that it is made of tiny colored pixels arranged on a grid.

Geras's code was expecting a pointillist pixel grid, but it was expecting a different *type* of pointillist pixel grid called a single-channel black and white image. In a single-channel black and white image, each pixel has only one value, from 0 to 255, where 0 is white and 255 is black. In my RGB image, each pixel had three values.

It is possible to do some math to clobber the RGB color image into a single black and white image. This is how the black and white photo filter on a phone or in an image editing program works. There are lots of options for math: you can average the colors or use complex algorithms to predict new colors. Robinson heroically tried several methods to convert the RGB images to single-channel black and white. Each one failed. The AI simply could not detect cancer in the screenshot images.

It was time to hunt down higher-resolution images. Again, I called the tech support people who run the medical record system and tried to escalate my request. At the highest level of tech support, I explained that I was trying to download high-resolution images of my 2019 mammogram.

"Which buttons did you click?" the tech support person asked. I explained that I clicked all the right buttons, and tried three different computers, two operating systems, and half a dozen different image-viewing software packages. Each time, I got an error: "This file is missing a DICOM attribute needed for processing." She and her colleagues couldn't figure out the problem. She suggested I request a CD. I was skeptical.

I tried the other hospital where I had received scans. I submitted an image request through the online portal, hoping I could get high-resolution images digitally. A few weeks later, when nothing came in via email or the portal, I called. The nice man on the phone said he could send me a CD of my images. This time, I was prepared.

"It's getting hard to find a CD player on a computer," I said. "Can you send me digital images?"

"We can't," he said. "We can only send a CD. If you were a doctor's office or an affiliated hospital, we could send it electronically using our system."

"What system is that?" I asked.

"Ambra," he said.

I looked it up quickly. It's a company that transfers medical image files. "Can I sign up for Ambra?" I asked.

"No," he said.

"Can I have the images sent to my doctor?" I asked.

"Only if your doctor is in the network," he said.

"Is my doctor in the network?" I asked, knowing that the two hospitals are less than a mile away from each other.

"No," he said.

I sighed and asked him to mail the CD to me. I then purchased a CD drive to read the files. It felt like absurdist theater. This friction is one of the reasons I am not optimistic when people talk about the frictionless future world, full of promise and seamless technology. I had encountered broken technology and broken systems every time I wanted to do something nonstandard with my own medical data.

Once the CD arrived, I put it in the drive and shared the files with Robinson. I had him run the properly converted high-resolution black and white images through the detection code again. The code inked a red box around the area of concern. It correctly identified the area where my cancer was. Success! The AI told me I had cancer. The chance that the identified area was malignant, however, seemed very low. The system generates two scores, one for benign and one for malignant, each on a scale of zero to one. The malignant score for my left breast was 0.213 out of 1. Did that mean there only a 20 percent chance that the image showed cancer?

I set up a video call with Kryztof Geras, my neighbor and the author of the code I was using. "That's really high," said Geras when I told him my score. He looked worried.

"I did actually have cancer," I said. "I'm fine now."

"That's pretty good for my model!" Geras joked. He sounded relieved. "It's actually accurate, I guess. I was afraid that it gave you a false positive, and you didn't have cancer." This was not a situation that any data scientist is prepared for: someone calling up and saying that they ran their own scans through your cancer detection AI. In theory, the whole reason to do open science is so that other people can replicate or challenge your scientific results. In practice, people rarely look at each other's research code. Nor, perhaps, do they expect quite such closeness between the data and their neighbors.

Geras explained that the score does not show a percentage, that he intended it to be merely a score on a zero-to-one scale. As with any such scoring system, humans would determine the threshold for concern. He didn't remember what the threshold was, but it was lower than 0.2. At first, I thought it was strange that the number would be represented as an arbitrary scale, not a percentage. It seemed like it would be more helpful for the program to output a statement like, "There is a 20 percent chance there is malignancy inside the red box drawn on this image." Then I realized: the context matters. Medicine is an area with a lot of lawsuits and a lot of liability. An obstetrician, for example, can be sued for birth trauma until a child is twenty-one years old. If a program claims there is a 20 percent chance of cancer in a certain area, and the diagnosis is wrong, the program or its creator or its hospital or its funder might be open to legal liability. An arbitrary scale seems more scientific than diagnostic, and thus it is less malpractice-attracting in the research phase.

As a lifelong overachiever, I was a little put out. It seemed like 0.2 out of 1 was a low score. I somehow expected that my cancer would have a high score. It was, after all, cancer—a thing that could kill me, and a common killer that had already killed my mother, a number of my family members, and several friends.

The difference between how the computer ranked my cancer and how my doctor diagnosed the severity of my cancer has to do with what brains are good at, and what computers are good at. We tend to attribute human-like characteristics to computers, and we have named computational processes after brain processes, but when it comes right down to it a computer is not a brain. Computational neural nets are named after neural processes because the people who picked the name imagined that brains work in a certain way. They were wrong on many levels. The brain is more than merely a machine, and neuroscience is one of the fields where we know a lot but essential mysteries remain. However, the name "neural nets" stuck.

The human brain excels at detecting anomalies. My favorite unscientific way of describing this is the tiger in the grass effect. Back in cave-dwelling days, let's say you were out gathering food near a stand of tall, waving grass. The grass is all uniform and shifts in the breeze. However, you notice something off about the way a patch of grass is moving. It's a tiger, in the grass, lying in wait so it can attack and eat you.

If you notice the anomalous movement of the grass, you can make noise and scare off the tiger, or get out your spear and stab the tiger and eat it, or whatever Paleolithic reaction you want. Basically, humans evolved to see anomalies for our own protection. The cave dwellers who didn't see the tiger got eaten; the cave dwellers who did see the tiger survived. The ones who survived passed down their anomaly-detection skills and the genetic propensity to see anomalies.

That ability to detect anomalies is at the heart of my doctor's ability to spot the malignant particles in an X-rayed blob. They trained in what different kinds of malignancies look like, they look at dozens of these things every day, and they are an expert in spotting cancer. A computer works differently. A computer can't instinctively detect something that is "off," because it has no instincts. Computer vision is a mathematical process based on a grid. The digital mammogram image is a grid, with fixed boundaries and a certain pixel density. Each pixel has a set of numerical values that represent its position in the grid and a color; a collection of pixels together makes up a shape. Each shape has a measurement of distance from the other shapes in the grid, and these measurements are used to calculate the likelihood that one of the shapes is malignant. It's math, not survival instinct. And survival instinct is one of the strongest forces in existence. It's also a little mysterious, which is okay too. We'll understand more and more every year as science and anthropology and sociology and all the other disciplines progress.

Geras and his coauthors are very specific in the paper about how good (or not good) their algorithm is, which is impeccable scientific practice. The authors write:

> On the random subset of data used in our reader study . . . a committee of radiologists achieved the macUAC of 0.704, while our model achieved the macUAC of 0.688. We conclude from these results that predicting BI-RADS without prior exams and information about the patient is very difficult even for well-trained human experts. Our neural network is already performing well in comparison. It is interesting to note that our model is clearly worse than the committee of the radiologists in recognizing BI-RADS 0 and clearly better in recognizing BI-RADS 2.[5]

Translation: predicting cancer from images alone is very hard for both humans and computers. Exams and other diagnostic information are essential. The BI-RADS scale the authors refer to is the Breast Imaging Reporting and Data System, which is a set of categories for talking about mammography results. A BI-RADS score of 0 means the mammography findings are incomplete, and follow-up imaging or examination is needed. A BI-RADS score of 2 means that benign findings were detected, like benign calcifications or lumps that are not of concern. A BI-RADS score of 5 means biopsy is recommended, and a score of 6 means there is cancer that has been validated by a biopsy. The authors are saying their AI is worse than a group of radiologists at recognizing situations that need follow-up. Their AI is better than a group of radiologists at recognizing findings that have been identified as benign. All things considered, they conclude that the AI is pretty good in lab conditions. I agree.

Here's something interesting about automated medical diagnosis. When a researcher like Geras builds an AI system, they have to choose how to calibrate the system mathematically. Every system is going to be wrong; the researcher has to decide how wrong, and what kind of wrong is preferred. Mathematically speaking, there are going to be false results because the results are set to fit a curve that they plot. The researcher has to make a decision: Is it better to have more false negatives or more false positives? A false negative would mean that the machine says there is likely no cancer, when there really is cancer. A false positive would mean that the machine says there may be cancer, and a human needs to look at the results and run more tests.[6] Every AI-based medical diagnosis system has to be calibrated in this way.

"In general, the cost of false negatives in medicine is much higher," Geras told me. "It is usually considered worse to miss someone who actually has a certain medical condition than to continue the imaging pipeline for someone who at the end of the day turned out to be okay." Most AI diagnostic systems are tuned to yield more false positives because this is the consensus in the community. This is a trade-off that is impossible to code our way out of. Unlike with human diagnosis, with machines you have to choose what kind of wrong diagnoses you prefer. It's more expensive to have more false positives and do more tests because a machine diagnosis sends up an alert, but it's preferable. Missing a cancer diagnosis can literally cost someone their life.

Smart people disagree about the future of AI diagnosis and its potential. I remain skeptical that this or any AI could work well enough outside highly constrained circumstances to replace physicians, however. Someday? Maybe. Soon? Unlikely. As I found in my own inquiry, machine learning models tend to perform well in lab situations and deteriorate dramatically outside the lab. Another reason I'm skeptical about AI diagnostics is based on the fact that machine learning models tend to be trained on data from a single clinical site, whereas they need to be tested on multiple clinical sites. In a 2021 study, researchers trained three different deep learning models on data from three different hospitals: the National Institutes of Health Clinical Center in Bethesda, Maryland; Stanford Health Care in Palo Alto, California; and Beth Israel Deaconess Medical Center in Boston, Massachusetts.[7] The datasets were chest X-rays, and the models gave a binary prediction for pneumothorax, or collapsed lung. The researchers chose pneumothorax because there are four medical devices cleared by the FDA to triage X-ray images to diagnose pneumothorax. There are also lots of publicly available pneumothorax datasets for testing.

The researchers found that each model performed well on the hospital data that it was trained on. The model trained on the NIH data performed well on NIH scans. However, when they fed the Stanford hospital data to the NIH model, the results went haywire. "Across the board, we found substantial drop-offs in model performance when the models were evaluated on a different site," the researchers wrote in *Nature*.[8]

Even worse, the performance disparity between Black and white patients increased when a model trained at one site was used on data from a different site. Single-site studies for testing are problematic, the authors write. Even though the FDA has approved many

AI-based medical devices tested at single sites, the AI diagnostic performance is not reliable outside the home site.

In places where diagnostic AI has been implemented, radiologists are not necessarily using it. In one study, researchers looked at the uptake of AI among diagnostic radiologists making breast cancer, lung cancer, and bone age determinations. The AI results were made available after the radiologist looked at the scans and made their assessment. The doctors looking at breast cancer and bone age consistently ignored the AI results, calling them opaque and unhelpful. Because the AI flagged so many conditions that the doctors ultimately labeled benign, the AI was just a hindrance and a tax on their time. The lung cancer diagnosticians, on the other hand, enjoyed using the AI most of the time. They used it to validate their results, experiencing pleasure and validation when the AI agreed with their assessment of lung scans.[9]

The doctors in the study all struggled with the lack of information given by the AI. The program delivered a result, without giving a human-comprehensible reason. An AI model fails silently, Geras says. There might be something weird happening with the model, or there might not be. We don't know. An AI model does not explain itself. If I had the original data, and the staff and the money and the time, I could go in and audit the dataset and figure out if text markings on my scans, for example, or some other factor was making a difference in the diagnosis. I don't have any of those things. It is an unknown. There are known unknowns in machine learning, then there are the unknown unknowns.

Most people in machine learning know about the common pitfalls, and they tell the same cautionary stories. There's one about a machine learning classifier that was fed photos of dogs, with the goal of identifying which of the photos were of huskies. It seemed to do exceptionally well at identifying which dogs were huskies—until someone realized that all the husky photos involved snow, and the classifier was detecting the presence of snow rather than dog features. There's another story about an AI that was excellent at detecting from X-rays which patient records would also have a diagnosis of pneumothorax—until someone realized that all of the X-rays in the training data included a chest tube, which had been put in because the patient's lung had collapsed. These cautionary tales help to explain why an AI trained on data from one hospital might not work on data from a different site. AI detects patterns in images that humans don't see. The AI might be detecting a pattern of the hospital name printed on the corner of the image, as well as the presence of a notable shape. The researcher is asking the AI to look for a notable shape (the cancerous tissue), but if the AI has trained itself to look for a notable shape plus the hospital name, it might not alert if given an image with the hospital name in a different location. Because an AI fails silently and does not give a human-comprehensible reason for its decision, a researcher may have trouble explaining why an AI works effectively on diagnostic images from one site and not from another.

People who have an economic stake in the game tend to be more tolerant of risk and uncertainty than those of us without an economic stake in the game. The people who work at medical device manufacturer ICAD, for example, are extremely optimistic about the future of AI diagnostics. The risk to their business is substantial, as they say in their investor report. I am a person with a very small risk of recurrent cancer. I am more reluctant to put my faith in AI cancer diagnosis.

One of the big things working against AI in radiology is money. Hospitals get paid by insurance companies when a radiologist reads a scan. Hospitals don't get paid by insurance companies when an AI reads a scan. There's a huge amount of money being spent here for an imaginary world in which, say, remote rural villages can get mammography machines and have an AI read a scan and then the person can know if they have breast cancer. But in remote rural villages, electricity is often a problem, as is internet connectivity and cell reception, and the air conditioning required to run computers. There's no guarantee that a mammography machine and AI radiology program would be able to function in a remote rural village. When the electricity is intermittent and there's no mail delivery to get the images on CD from one place to another, AI is hard to do. When the machine breaks, who will fix it? Who will pay for the regular equipment updates? If the AI detects cancer, who will do the surgery and pay for the hospital care to treat the patients and provide physical therapy afterward? The fantasy of computerized medicine sounds a lot like the fantasy of the self-driving car: fascinating, but impractical. In fact, a number of studies suggest that breast self-exam is a more useful strategy than mammography for remote places, because of the financial, logistical, and sociocultural constraints. In sub-Saharan Africa, for example, there is often a delay between when someone detects a lump and when they seek healthcare, and routine mammograms are not a priority for many.[10] Much of the public health literature suggests that reducing cancer mortality in countries throughout the Global South starts with low-tech screenings and getting people access to medical care—not necessarily using the most high-tech equipment and methods.[11] The huge amount of money being poured into AI diagnostics, when simpler methods could have a high impact, is an example of technochauvinism.

Machine models are insufficiently flexible when things change. A good example of this is what happened with mammograms after COVID-19 vaccines. In the months after the vaccines became widely available, radiologists started seeing scores of women with enlarged lymph nodes. There was a flood of women who had put off routine mammograms, and they were catching up on routine medical care that had been postponed by pandemic lockdowns. They were also coming to radiologists because they were concerned about large lumps found in their armpits. Doctors quickly realized that the lumps were a temporary condition brought on by vaccination. The COVID vaccine led to temporarily enlarged lymph nodes as part of a normal immune response. A machine learning system would not have been able to look at enlarged lymph nodes and connect this to COVID vaccine response—because COVID was a brand-new condition, first seen after the machine learning models were trained. Models can't update their frames of reference as quickly or as flexibly as human experts.

Models can be used quite effectively to distribute information, however. I found out about the post-vaccine enlarged lymph nodes because of an article that was suggested to me by the recommendation engine on the *New York Times* site, an engine that uses AI. Months later, after I got a COVID-19 booster shot, I developed a large lump under my arm. I knew not to freak out and think I had cancer again—because I had read an article written by a person and distributed by an AI.

10

Creating Public Interest Technology

Thus far, we've talked about problems. We've discussed problems that exist in the world, and the ways they are reproduced inside computational systems. We've also discussed ways to spot the problems. It's time to move on to the part everyone likes: the hopeful part.

We've become conditioned to look for the hopeful solution at the end of every story. This is the idea behind solutions journalism, which aims to tell stories about bad things in the world along with proposed solutions for the horror. People who are accustomed to this narrative often expect that there is an easy answer to computational problems. In tech, many have become addicted to what we might call the "lean startup" pattern, which goes something like this:

- Identify a pain point.
- Narrow down the pain point to something we can write code against.
- Identify a target market who can use this solution.
- Write the code, solve the problem.
- Sell the code to the target market.
- Scale up until you take over the world.

Nobody except Big Tech firms actually gets to the last step (and whether this has helped or hurt our society is an ongoing debate in any case). But even if we could make this narrative a reality, it doesn't really work for complicated problems. This method works for simple problems only.

For the complicated problems, a better frame is that of public interest technology. Public interest technology is exactly what it sounds like: creating technology that is in the public interest. Sometimes public interest technology means building better government technology or policy. Sometimes it means doing projects that hold algorithms and their creators accountable. "We define public interest technology as the application of design, data, and delivery to advance the public interest and promote the public good in the digital age," write Tara Dawson McGuinness and Hana Schank in their book *Power to the Public*.[1] Algorithmic auditing and algorithmic accountability reporting are two strains of public interest technology that I think show the most promise for remedying algorithmic harms along the axes of race, gender, and ability. I'm optimistic about public interest technology because the people involved in it are centering the public good and using technology in order to advance collective well-being, instead of centering profit and ignoring systemic inequality.

Public interest technology is a new field that developed in the period of 2013–2015 following President Barack Obama's reelection campaign and the launch of HealthCare.gov. Obama's 2012 campaign was the most high-tech political campaign that had ever been done at that point. It relied heavily on data, social media, and digital tools. After the campaign, however, the people who so skillfully deployed data and technology in the campaign did not transition into public service jobs as campaign workers had in the past. They went to the private sector, many to Silicon Valley. The Ford Foundation and the MacArthur Foundation noticed this shift, and became concerned about what it meant for the pipeline of government talent.

As technology became more essential to everyday life, who would develop tech policy and help government software systems evolve if talent was flowing away from the public sector?

The foundations' concern was magnified after the disastrous 2013 launch of HealthCare.gov, the site that was supposed to support the 2010 Affordable Care Act, also known as Obamacare. The site was supposed to be a clearinghouse that would let Americans compare prices on healthcare plans, find out if they were eligible for government healthcare subsidies, and secure healthcare coverage. Outsourced government contractors built the site, and most of the things that could go wrong did. The site crashed, the code was buggy, the servers were not robust enough to withstand the millions of people who tried to use the site simultaneously. It was an epic fail. The Obama administration assembled a strike team to help address the situation. The team secured new contractors, the site was fixed and redesigned, better management processes were put in place, and the crisis passed within a few months. Nobody has heard much about HealthCare.gov since its resurrection. "That's what you want in a government website," points out Cori Zarek, executive director of the Beeck Center for Social Impact + Innovation at Georgetown University, who was a White House tech advisor at the time. "You want it to just work, and you don't want it to be in the news."

The HealthCare.gov episode made it clear that the US government needed to adapt its processes and increase its in-house tech talent in order to better manage government software, create and enact effective tech policy, and use digital tools to protect citizens' interests. By 2015, this effort became known as public interest technology. Public interest tech was founded for many of the same reasons that the field of public interest law was founded in the 1960s: to get more talented people working on projects to serve the public good. It derives from previous movements including civic tech, which had very similar concerns and led to lots of new initiatives around transparency and open government data.

Inside the government, public interest technologists are building new, more efficient software systems and are updating creaky, out-of-date government processes. "I think it takes a while to realize procurement and hiring are not just boring admin tasks, but central to how everything happens," Hana Schank said in an interview with the US Digital Service (USDS). USDS is a tech startup working across the federal government to deliver better services to the American people. Change has been slow, but with groups like USDS and 18F, a digital services agency within the US government, bringing the dynamism of civic tech to Washington, effective modernization is happening. "You don't just show up one day and say, hey you should have some metrics in place, or hey you should talk to some users, and expect agencies that have had a [standard operating procedure] in place for 100 years to go, 'oh gee that's a great idea, we'll get on that,'" Schank said of the pace of change. "But I think there is more pressure building from the outside world, and more expectations around how government should function, that will help encourage those changes. So when those changes do happen more globally throughout government, it will be in part because USDS has been there on the inside saying the things that people outside of government are now starting to say."[2]

Outside of government, a developing area of public interest technology called algorithmic auditing shows great promise for decreasing bias and fixing or preventing algorithmic harms. Algorithmic auditing is the process of examining an algorithm for bias or unfairness, then evaluating and revising it to make it better. Algorithms are always examined in context because they are as social as they are technical. Rarely are the problems solved in one shot. But auditing is the best thing we have.

Auditing comes from the compliance world. If you're not familiar with compliance, you are lucky! It is the norm in highly regulated industries like finance or insurance. It's a normal process to make sure that everything is working okay and is not breaking any laws. It ensures that people are doing their jobs and checking on the things that need to be checked on, so we can prevent disasters like the 2008 financial crisis.

Think about the spot in your house that descends into chaos without constant attention. Maybe it's your closet, maybe it's the spot where everyone puts keys and empties out their pockets. There's always a spot (possibly many) that collects chaos. In my apartment, it's the end of the kitchen counter nearest to the front door. It's a fact in the real world, and it's a fact in the virtual world too—people forget to do things, and there are digital spots that collect chaos. In the real world, you need to regularly clean and organize the chaotic spot. Compliance (in an ideal world) is the business process that reminds you and requires you to organize the ordinary chaos of doing business while complying with the law. Compliance processes often exist because legal problems have happened in the past, and people are trying to prevent similar problems in the future; the processes are slightly unpleasant but crucial. Compliance means that things don't fall through the cracks, that civil rights are not violated, and that companies are following the law.

Algorithmic auditing means looking at an algorithm to see what it is doing and what are the possible points of failure or bias. Ideally, algorithmic auditing will be adopted as part of the compliance process for a range of industries. It hasn't been adopted widely in the United States yet, but EU regulatory progress like General Data Protection Regulation (GDPR) and the 2021 EU proposed AI legislation suggest that compliance for AI is coming soon. As far as the question of why audit at all, I think it is best articulated by auditing expert Inioluwa Deborah Raji, who tweeted: "We can't keep regulating AI as if it works. Most policy interventions start with the assumption that the technology lives up to its claims of performance, but policymakers & critical scholars need to stop falling for the corporate hype and should scrutinize these claims more."[3]

The laws and policies that algorithms must follow differ by industry and geographic area. In the United States, the Federal Trade Commission (FTC) points out three of the many laws that are relevant to AI developers: Section 5 of the FTC Act, the Fair Credit Reporting Act, and the Equal Credit Opportunity Act. In an April 2021 guidance, the FTC explains:

> **Section 5 of the FTC Act.** The FTC Act prohibits unfair or deceptive practices. That would include the sale or use of—for example— racially biased algorithms.
> **Fair Credit Reporting Act.** The FCRA comes into play in certain circumstances where an algorithm is used to deny people employment, housing, credit, insurance, or other benefits.
> **Equal Credit Opportunity Act.** The ECOA makes it illegal for a company to use a biased algorithm that results in credit discrimination on the basis of race, color, religion, national origin, sex, marital status, age, or because a person receives public assistance.[4]

Until recently, software developers have not paid enough attention to ensuring their algorithms operate within existing laws. Auditing is a way to make sure that the public interest is being preserved in and around algorithms. Generally, we think about two ways of auditing: bespoke and automated. In bespoke auditing, we do an audit by hand: we break down the process and write documents and have meetings. In automated auditing, we do the same thing and use additional technical components to analyze the performance of a system on the level of code, using a platform or repeated tests. There are more thresholds in the automated method. Currently, there are about twenty-one different mathematical definitions of fairness.[5] Interestingly, these definitions are mutually exclusive.[6] It is mathematically unlikely that any solution can satisfy one kind of fairness, and also satisfy a second criteria for fairness. So, in order to consider an algorithm fair, a choice will have to be made as to which kind of fairness is the standard for this type of algorithm. From a policy perspective, this means that all similar algorithms would need to be evaluated according to the same fairness metric.

One of the people leading the field in algorithmic auditing is Cathy O'Neil, the author of *Weapons of Math Destruction*. Her book is one of the catalysts for the entire movement for algorithmic accountability. O'Neil's consulting company, O'Neil Risk Consulting & Algorithmic Auditing (ORCAA), does bespoke auditing to help companies and organizations manage and audit their algorithmic risks. I have had the good fortune to consult with ORCAA. When ORCAA considers an algorithm, they start by asking two questions:

• What does it mean for this algorithm to work?
• How could this algorithm fail, and for whom?

One thing ORCAA does is what's called an internal audit, which means they ask these questions directly of companies and other organizations, focusing on algorithms as they are used in specific contexts. They have also asked these starting questions of regulators and lawmakers in the course of developing standards for algorithmic auditing. ORCAA's approach is inclusive: they aim to incorporate and address concerns from all the stakeholders in an algorithm, not just those who built or deployed it. It is essential to include members of an affected community in an audit in order to evaluate whether harms have occurred.

ORCAA has worked with computer scientist Joy Buolamwini's organization, the Algorithmic Justice League, to perform audits with an intersectional focus. In Buolamwini's paper "Gender Shades," she and her coauthors propose an intersectional framework for analyzing an algorithm.[7] This means evaluating the algorithm's performance for different subgroups. Not just men and women, but perhaps also nonbinary and trans folks, and for darker-skinned women and lighter-skinned men. Intersectionality looks at the intersection of different groups that an individual belongs to, like race and gender, and proposes that the intersection gives rise to different experiences and different forms of oppression or discrimination. The matrix of domination is different depending on whether you're a Latinx trans woman or a Black man or an Afro-Caribbean woman or a Pacific Islander domestic worker or a disabled Native

American CEO or any other combination of identities. This framework is helpful for identifying for whom an algorithm could fail. When you start thinking about race and gender and ability explicitly, and writing down an intersectional matrix of people for whom the algorithm might fail, it becomes easier to spot problems.

Auditing is fascinating because you get to dig into the history of an algorithm and weigh competing corporate and mathematical imperatives. In auditing, one of the things we do is translate extremely complicated mathematical concepts for different corporate audiences.[8] In math, you look to prove theorems that hold true everywhere, across time and space, in the same way. This is what physicists do too, but for the natural world. People trained in math and physics (which includes many data scientists and computer scientists) often make predictable mistakes when writing code for social contexts because they are looking for one method that explains everything. In auditing, there's less of a focus on one single explanation, and we consider both quantitative and qualitative factors.

Auditing involves a lot of creativity, looking for the edge cases and figuring out what could go wrong. We look at the code or the design pattern that went into making an algorithmic system, and do what in other contexts might be called threat modeling. Ruha Benjamin's idea of tech discrimination is my own animating principle.[9] Benjamin offers the frame that technology discriminates by default, not that discrimination is a glitch. Adopting this point of view makes it easier to see where technology might be disadvantaging certain groups or violating people's civil rights. My other default assumption is that AI doesn't work as well as people imagine. This perspective also makes it easier to spot algorithmic problems.

In addition to internal audits, there are external audits, which (as the name suggests) are performed outside the company without access to code or trade secrets. Usually, external audits are initiated by journalists, lawyers, or watchdog groups. ORCAA, for example, has helped attorneys general to identify and prosecute cases where algorithms are used to break the law. An attorney general has power to demand (via subpoena) documentation, system data, or code from the target company. External projects are sometimes quite creative. For example, a watchdog project called Exposing.AI lets you find out if your photos were used to train facial recognition systems. This is more common than most people expect. People post pictures online imagining they will be shared with their friends or their audiences. Rarely are they excited to find that their photos have been collected and used to train AI models. When the news broke that Clearview AI had scraped millions of Flickr photos and used them to create a facial recognition database for policing, there was a massive outcry. Clearview AI argued that its use of Flickr images was within the labeled use of the images. People do not feel as if it was an ethical use of their images, however.[10]

When journalists perform an audit in order to hold an algorithmic system accountable, it is called algorithmic accountability reporting. Julia Angwin's "Machine Bias" investigation for ProPublica, which revealed the COMPAS algorithm is biased, is one of the first and best-known. Investigative journalists have less access than lawyers or internal auditors, but they often find innovative ways to access the necessary information and draw conclusions.

Julia Angwin went on to found The Markup, the news organization I've mentioned several times, which is currently doing some of the world's most cutting-edge algorithmic accountability reporting. The Markup's slogan is "Big Tech is watching you. We're watching Big Tech." Their investigations have resulted in numerous positive changes since they launched in 2020. An August 2020 investigation found that state coronavirus information web pages had serious accessibility issues that would prevent disabled users from accessing crucial pandemic information. "Black-box underwriting decisions are not necessarily creating a more level playing field and actually may be exacerbating the biases that feed into them," said Rohit Chopra, the director of the Consumer Financial Protection Bureau, who launched a new effort to combat discriminatory mortgage lending in response to The Markup's findings about bias in mortgage-approval algorithms.[11]

The newsroom often fact-checks claims made by Big Tech firms. Social media companies, for example, have been caught making false claims and enabling illegal behavior on their platforms. "Facebook's vice president of content policy, Monika Bickert, testified that Facebook doesn't allow discriminatory targeting for 'certain types of advertisements, such as financial services' during a Senate Judiciary subcommittee hearing on April 27 [2021]," wrote The Markup reporter Albert Ng. "Two days later, The Markup published an investigation showing Facebook was doing just that. We found several ads for credit cards and other financial services targeted by age."[12] As of this writing, Congress is investigating. Although Facebook policy prohibits ads for job opportunities, housing, and certain financial services from targeting by age, race, gender, zip code, sexual orientation, gender identity, family status, disability, and medical conditions, a number of investigations by The Markup have shown that this policy is easily flaunted. In some cases, such targeting violates civil rights law. Other investigations have been cited in antitrust actions against Google and have questioned whether Amazon and Google give preference their own products in search results, thereby undermining a free, open, competitive marketplace.

One of the advantages to The Markup's method is that citizens can replicate the newsroom's investigative methods. Most stories come with an extensive methodology section explaining how the investigation was performed and validated. This philosophy of "show your work" increases transparency and allows the audience to trust the newsroom's findings.

Another external audit by academic researchers at University of California, Berkeley, revealed that Black and Latinx people pay more for mortgages and are denied at higher rates. "Latinos and

African-Americans paid almost one tenth of a percentage point more for mortgages between 2008 and 2015, the study found—a disparity that sucked hundreds of millions of dollars from minority homeowners every year," wrote CBS reporter Kristopher J. Brooks about the study.[13] Black and Latinx borrowers end up paying an additional $765 million per year in additional mortgage costs—a disparity that contributes to the racial wealth gap. Over time, if models were allowed to continue this discriminatory pattern, the racial wealth gap would become insurmountable.

Reporting based on whistleblower testimony and leaked documents is also a kind of algorithmic accountability reporting. For example, the *Wall Street Journal*'s 2021 Facebook Files project was based on documents provided by Frances Haugen, a former Facebook employee. The documents and reporting showed that the company was aware of how its algorithms were causing ill effects and spreading hate and misinformation. "Time and again, the documents show, Facebook's researchers have identified the platform's ill effects," wrote the reporters. "Time and again, despite congressional hearings, its own pledges and numerous media exposés, the company didn't fix them. The documents offer perhaps the clearest picture thus far of how broadly Facebook's problems are known inside the company, up to the chief executive himself."[14] Other findings of the Facebook Files project were either shocking or confirmed what had been rumored for years. Instagram's algorithms were known to be especially toxic for teenage girls, leading them to eating disorder content and having negative impact on their mental health. Drug cartels and human traffickers were identified on the platform, but employees found the company responded inadequately or not at all. Facebook spread a flood of antivaccine content during the COVID-19 pandemic. In response, Facebook changed its parent company's name to Meta, in an apparent attempt to distract public attention. Another report by the *Washington Post*, based on leaked documents, showed that Facebook spread nationalist propaganda, anti-Muslim hate, and misinformation in India and "was well aware that weaker moderation in non-English-speaking countries leaves the platform vulnerable to abuse by bad actors and authoritarian regimes."[15] As of this writing, there is movement in Washington to rein in Big Tech and make policy changes that it is hoped will improve the public conversation and intervene in both present and future algorithmic harms.

Auditing, especially automated auditing, is important because models decay. Many people imagine that they will be able to create or implement a computational system and then "set it and forget it." Nothing could be further from the truth. Every computational system needs to be updated, staffed, and tended. Computer systems need to change as the world changes. Remember that an algorithmic system is made up of multiple components. We refer to it as an algorithmic system, but really there are multiple interlocking systems and models and rule sets.

Figuring out which fairness metrics to use is one of the biggest auditing challenges. We need to audit algorithmic systems for search, e-commerce, online advertising, ad tech, maps, ridesharing, online reviews and ratings, natural language processing, ed tech, recommendation systems, facial recognition inside and outside policing, predictive policing, criminal justice, housing, credit, background-checking, financial services, insurance, child protective services, and more. These systems all operate in different contexts, and the same test won't necessarily suit every industry. An auditor needs to choose which single fairness metric works best for each algorithm in each specific context. She can then choose among multiple software packages to run the chosen fairness test. One popular open-source package is called AI Fairness 360. Originally developed by IBM, it is now managed by the Linux Foundation, which also oversees the Linux operating system.[16] AI Fairness 360 includes ten bias mitigation algorithms, including optimized preprocessing, reweighing, adversarial de-biasing, reject option classification, disparate impact remover, learning fair representations, equalized odds post-processing, and meta-fair classifier. There are also platforms for auditing, such as Parity and Aequitas. I am helping ORCAA to build a system called Pilot for automated algorithmic auditing.

A few other high-profile audits are instructive about the benefits of auditing. The "Gender Shades" project was obviously an audit. ORCAA audited HireVue, a company that sells algorithmic hiring software. Afterward, HireVue announced it was dropping its most controversial component, video analysis. Christo Wilson, a computer scientist at Northeastern University, and colleagues collaborated with Pymetrics, another HR firm, to perform an audit. This was notable because the Wilson team began the academic paper about the audit with a statement that undermines the practice of algorithmic resume-sorting, which is the basis of Pymetrics' business model.[17] They write that "there is no reason to assume a priori that ML [machine learning] systems in the hiring domain will automatically be 'objective,' 'neutral,' or 'bias-free.' Indeed, algorithm audits of gig-economy marketplaces and traditional resume boards have uncovered race and gender biases in these systems."

Microsoft researchers developed what they call a "community jury," a process for soliciting input from stakeholders during the software development process. Community input is not a new idea, but embedding it in business processes as a step to stave off algorithmic harms is a new development.

Auditing is not a silver bullet, it is a tool in an imperfect system. It does not always work. One failure of auditing happened in New York City, when the city made a task force devoted to cataloguing and overseeing all of the city's algorithms. They started by trying to develop a list of all the algorithms used by the city. The task force, which was made of multiple people with world-class reputations in algorithmic fairness, disbanded after only a year. It didn't have enough funding or resources, and the city didn't have the capacity to do what it said it was going to do with the task force. "The task force was given no details into how even the simplest of automated decision systems worked," wrote task force member Albert Fox Cahn, founder and executive director of the Surveillance Technology Oversight Project, about the fiasco. "By January 2019, there was growing anger about the city's unwillingness to provide information on what automated decision systems it already used. This undercut the value of the task force, which aimed to escape the theories and generalizations of the ivory tower to examine how these tools were operating in the real world, using the country's largest city as our test case. Only we never got the data."[18]

The effort needed to be better resourced, have more power to compel disclosure, and needed far more time than it was given. Auditing is not inexpensive and needs wide-ranging institutional support. The NYC algorithm task force disaster emphasizes that it's important to have the ability to say "no" to the tech if it is not working well or as expected. Few people are prepared to let software projects go; it's hard to kill your darlings after you've written them, and it's even harder to kill your darlings after you've invested thousands or millions into developing them. Everyone making an algorithmic system needs to be prepared to confront the shortcomings of the computational system and of the larger sociocultural context.

Another instructive audit situation comes from STOP LAPD Spying Coalition, a grassroots organization that led demands to audit PredPol, a predictive policing system used by the LA Police Department. "We released a report, 'Before the Bullet Hits the Body,' in May 2018 on predictive policing in Los Angeles, which led to the city of Los Angeles holding public hearings on data-driven policing, which were the first of their kind in the country," said the organization's founder, Hamid Khan. "We demanded a forensic audit of PredPol by the inspector general. In March 2019, the inspector general released the audit and it said that we cannot even audit PredPol because it's just not possible. It's so, so complicated."[19]

This was interesting to me because it gets at some of the essential problems of auditing software. A forensic audit is different from an algorithmic audit, and not everyone in the legal or forensic world knows that algorithmic auditing exists. It's likely that the inspector general's office didn't understand how an algorithmic audit would work, and thus claimed it was impossible. An audit is possible, yes— but it requires the inspector general's office and the auditors and the audit report-readers and everyone else in the institutional context to have a level of mathematical and computational literacy in order to understand and communicate about the results. When you ask for medical test results, you get a report, and lawyers and activists understand that. The artifact of the test results is a normal part of the discovery and legal sense-making process. When you ask for algorithmic test results, few people know what that means. There isn't yet a standard report. An algorithm is itself a kind of amorphous thing in most contexts. When you look at an algorithmic system, or a machine learning model, it looks like gobbledygook unless you can read code and understand data and know the right kind of reports to request from exactly the right person. It's confusing.

Responsible AI can be embedded in ordinary business processes in order to address ethical concerns. A good organizational process is outlined in the following list:

- Make an inventory of algorithms and algorithmic system vendors used by a company.
- Audit one algorithm.
- Remediate any harms from that algorithm and reshape business processes to prevent these harms in the future.
- Learn from the process to proactively seek out other places in the company where similar problems could occur.
- Update business processes to include ongoing algorithmic review.
- Audit and remediate more algorithms, repeating the steps above as many times as necessary.

Auditing doesn't have as much marketing hype behind it as innovation does. In part, this is because there is lots of funding (venture capital and otherwise) for building new things, but very little funding for fixing and improving the things that already exist. Public interest technology pushes back against this, with an awareness that we need to fund infrastructure as well as innovation. If we are building AI systems that intervene in people's lives, we need to maintain and inspect and replace the systems the same way we maintain and inspect and replace bridges and roads. Another thing that will help is decoupling innovation from social progress. Innovation and social progress are *not* the same thing. Using more technology does not bring about social progress if the technology causes algorithmic harms or (as is often the case) reverses hard-won civil rights advances. Finally, diversifying the landscape of technology creators will help, so that there are more people in the room who can bring more viewpoints and can raise awareness of potential issues that will need to be audited.

11
Potential Reboot

Since I started publishing about issues of algorithmic accountability and algorithmic justice in 2017, there has been a wholesale revolution in how people think about these issues. Safiya Umoja Noble told me that when she was in graduate school, drafting the work that became her groundbreaking 2018 book *Algorithms of Oppression*, people thought she was crazy for writing about how technology could be racist. Similarly, Shalini Kantayyaa, the director of the documentary *Coded Bias*, told me that for years she had trouble talking to people at parties because they would ask what she was working on and she would say "a film about racist robots" and they would look at her like she was nuts. Kantayyaa and I met at a dinner party and spent most of the time talking about racism and robots. For both of us, it was the first time we'd run into someone at a party who understood what we were concerned about. Now, Safiya Noble has been awarded a MacArthur "Genius Grant." Meghan Markle is reading *Algorithms of Oppression*; Joy Buolamwini is in the 2021 *Vogue* September issue as a model for Black women in STEM; and *Coded Bias* is streaming to millions on Netflix. For this amount of change to have happened in just five years is unprecedented. It bodes well for the future.

I hope this book has given you tools to look at a new situation and see how technochauvinism and bias may be at work. I also hope you feel empowered with enough computational literacy to call bullshit on bogus technological claims. I am optimistic that in the future, more people will become aware of how racism and gender bias and ableism permeate mainstream technology, and will work to eliminate these problems. A few notable developments in these areas suggest that the world is moving toward algorithmic justice.

For years, the National Football League has been using a race-based algorithm to determine payouts to football players who develop dementia or other brain injury complications. Black players, who make up 70 percent of the league, were denied benefits from the $1 billion in claims paid through a massive class action settlement resulting from a 2012 case brought against the NFL by former players. "The race-based norms that had been used in the N.F.L. dementia tests—one for white former players and another for Black former players—assumed that Black former players started with worse cognitive function than white former players," wrote Ken Belson in the *New York Times*.[1] Finally, in 2022, a federal judge threw out the race-based assessments on the grounds that they are discriminatory. This opens the door for Black former players to have their cases reevaluated and to potentially recover hundreds of thousands of dollars in compensation for job-related injuries. This could set a precedent for eliminating other race-based "corrections" in medicine.

In the policy world, I am enthusiastic about the 2021 proposed EU AI legislation. The European Union (EU) proposed a new piece of AI legislation that calls for high-risk AI to be controlled and regulated. The proposed legislation divides AI into high-risk and low-risk categories and specifies which kinds of AI fall into which category. Facial recognition in policing might be considered high-risk; unlocking your phone with your face might be low-risk. Under the proposed regulation, anyone using high-risk AI would be required to register it and demonstrate regularly that the AI is not being discriminatory.

This is absolutely a step in the right direction. The proposal calls for a ban on biometric identification in real-time feeds, which means the dream of real-time facial recognition for surveillance would be dead and surveillance capitalism would be thwarted in this one small part of the commercial marketplace. This is another win.

The EU legislation also suggests that firms should use a regulatory sandbox to test their algorithms for discrimination. A regulatory sandbox is a concept first popularized in financial technology. It's a virtual space where algorithms can be tested and evaluated—just like a kid can build a castle safely in a sandbox, then destroy the castle immediately afterward without consequences. A regulatory sandbox is a safe space for testing algorithms or policies before unleashing them on the world. I got very interested in this idea, and together with some colleagues at ORCAA I am currently building an auditing system, Pilot, that is based on the regulatory sandbox idea. We're designing Pilot to have dials and gauges that show intersectional analyses and allow compliance with the new proposed EU legislation as well as future US legislation. We're optimistic about the possibilities. If you are going to use an algorithmic system, you should be required to prove that it complies with the law of the land and does not discriminate. I am hopeful about other global changes like Amnesty International's proposal Ban the Scan, which would reduce or eliminate facial recognition surveillance in Hyderabad, India, one of the world's most-surveilled cities.

Legislative progress is happening in the United States as well. At least thirty bills were proposed in 2021–2022 to limit Big Tech. These bills include Section 230 reform, antitrust, and other measures. Particularly promising is the Algorithmic Accountability Act of 2022, which would require companies to evaluate algorithms for bias and for effectiveness. "It is long past time Congress act to hold companies and software developers accountable for their discrimination by automation," said Representative Yvette Clarke, one of the bill's sponsors. "With our renewed Algorithmic Accountability Act, large companies will no longer be able to turn a blind eye towards the deleterious impact of their automated systems, intended or not. We must ensure that our 21st Century technologies become tools of empowerment, rather than marginalization and seclusion."[2] Not all of this legislation will pass, but it's a good start. As US president, Joe Biden has assembled a tech policy team that has more socio-technical expertise than any previous administration. Alondra Nelson, a renowned social science scholar whose most recent book is *The Social Life of DNA: Race, Reparations, and Reconciliation after the Genome*, leads the White House Office of Science and Technology Policy. Lina Khan, a noted antitrust legal scholar, is chairing the US Federal Trade Commission (FTC). For those of us who believe Big Tech needs to be broken up, it is a good sign. Breaking up Big Tech is not the end of the world, as many people would claim. We have good models for how to do this in the trust-busting of the Progressive era, and more recently in the breakup of AT&T and Microsoft.

In April 2021, the FTC published guidance on corporate use of AI. "If a data set is missing information from particular populations, using that data to build an AI model may yield results that are unfair or inequitable to legally protected groups," reads the FTC guidance. "From the start, think about ways to improve your data set, design your model to account for data gaps, and—in light of any shortcomings—limit where or how you use the model."[3] Other tips include watching out for discriminatory outcomes, embracing transparency, telling the truth about where data comes from and how it is used, and not exaggerating an algorithm's capabilities. If a model causes more harm than good, FTC can challenge the model as unfair. This guidance put corporate America on alert. Companies need to hold themselves accountable for ensuring their algorithmic systems are not unfair, in order to avoid FTC enforcement penalties.

Social movements like Black Lives Matter and #MeToo have prompted tremendous social change in the past few years, and they have inspired similar movements toward racial and gender equity in the tech world as well. Joy Buolamwini's organization, the Algorithmic Justice League (AJL), is doing a project called CRASH, which stands for Community Reporting of Algorithmic System Harms. Part of the project involves establishing bug bounties. The concept of a bug bounty comes from cybersecurity, where people can be rewarded by tech companies for finding and reporting system bugs. Bounties arose because hackers were finding bugs in systems and holding companies' data for ransom. If a company proactively paid hackers to find security holes, they could patch them and didn't have to suffer through a ransom attack.

AJL is taking this concept of bug bounties and applying it to algorithmic systems. They are collecting stories of algorithmic harms, and also proactively testing systems to find algorithmic harms. The hope is to establish rewards for people who find algorithmic harms, so that companies can proactively address and remediate the harms.

The algorithmic bug bounty is becoming a popular strategy. In July 2021, Twitter's ML Ethics, Transparency, and Accountability (META) team under Rumman Chowdhury offered a hacking session at a security conference focused on finding algorithmic harms. Twitter presented a single algorithm to the attendees and offered bug bounties to teams who could find problems with it.

A variety of new groups are working for social change in tech. Policy groups like Data for Black Lives, led by Yeshimabeit Milner, are working hard to craft legislation that minimizes bias while raising awareness among a wide range of groups. AI for the People, a nonprofit communications agency led by Mutale Nkonde, is working to educate Black communities about the racial justice implications in tech and to empower them to make their voices heard. Data & Society and NYU's AI Now are think tanks publishing thoughtful work on a range of algorithmic justice issues. STOP LAPD Spying Coalition, the group that requested an LAPD audit, is an alliance of different organizations that come together to collaborate and take collective action together toward common goals, rejecting police oppression and policies that make all people suspects in the eyes of the state. Their vision is the dismantling of government-sanctioned spying and intelligence gathering, in all its multiple forms. The Surveillance Technology Oversight Project (S.T.O.P.) litigates and advocates for privacy, working to abolish local governments' systems of mass surveillance. S.T.O.P.'s work highlights the discriminatory impact of surveillance on Muslim Americans, immigrants, the LGBTQIA+ community, Indigenous peoples, and communities of color, particularly the unique trauma of anti-Black policing. S.T.O.P. has collaborated with Princeton University's Ida B. Wells Just Data Lab, NYU Law's Brennan Center for Justice, the Immigrant Defense Project, Brooklyn College's Policing & Social Justice Project, and LatinoJustice PRLDEF, which are just a few of the other organizations doing great work in this space.

For many years, Big Tech has avoided worker organizing by paying enormous salaries to a privileged corps of engineers and executives. The 2018 Google walkouts were early signs of the worker organizing movements. Now that Big Tech is such a large employer, the need for worker solidarity is becoming clear just as it has in other industries like coal mining or media. Research project Collective Action in Tech documented more than 230 workplace actions at tech companies from 2019 to 2020. By contrast, they documented only sixty-one actions in the three years prior. "One of the biggest reasons for this increase in organizing is the growing recognition by workers (and the public) that not all tech is ethical and that tech companies continue to demonstrate their inability to self-police their morality," writes UnionTrack, a firm that makes software for labor union organizing and management.[4] Whistleblowers like Frances Haugen, Sophie Zhang, Yael Eisenstat, and Roger McNamee have started a public conversation about the ways that Meta (formerly Facebook) and other social media platforms have misled their workers as well as the public.

Hashtag activism has been an important way of building enthusiasm for labor and social movements.[5] #AppleToo is a movement of tech workers who have protested Apple's culture of secrecy and its lack of public scrutiny, which have led to workplace injustice for historically marginalized groups of people. "When our stories are collected and presented together, they help expose persistent patterns of racism, sexism, inequity, discrimination, intimidation, suppression, coercion, abuse, unfair punishment, and unchecked privilege," the group writes.[6]

Policy changes and social actions have been accompanied by changes in the academy. The global community around the ACM Fairness, Accountability, and Transparency (FAccT) conference has helped initiate a variety of ethics courses at engineering schools and in computer science departments. Though it has been a recommended part of the core computer science curriculum for years, only in the past few years have schools started requiring a standalone ethics course or an ethics component of required computer science classes. Courses like security researcher Seny Kamara's "Algorithms for the People" at Brown University are becoming more popular and more common. The work being produced by the FAccT community has also prompted a new ethics component to the annual Conference on Neural Information Processing Systems (NeurIPS), the largest AI conference. In 2021, NeurIPS instituted an ethics review as part of its peer review process, in addition to content and technical reviews. This should help to prevent publication of, for example, phrenology and other pseudoscientific works. The Public Interest Technology University Network (PIT-UN), to which I belong and which has funded my work, is creating new pathways for research and careers in responsible tech and public interest technology.

In 2020, a group of civil society organizations issued a joint statement in support of a United Nations report entitled "Racial Discrimination and Emerging Digital Technologies: A Human Rights

Analysis," by E. Tendayi Achiume. Achiume is the UN Special Rapporteur on contemporary forms of racism, racial discrimination, xenophobia, and related intolerance.[7] The statement (which I cosigned) makes a set of bold declarations toward algorithmic justice. One such statement is: "Technologies that have had or will have significant racially discriminatory impacts should be banned outright." Another is: "Technologists cannot solve political, social, and economic problems without the input of domain experts and those personally impacted."[8] The UN's awareness of algorithmic justice issues is sure to amplify the effect.

UCLA's Center for Critical Internet Inquiry, which helped draft the statement for the UN, has another creative idea for Big Tech: reparations. "We need new paradigms, not more new tech," wrote the center's director, Safiya Umoja Noble. "We need fair and equitable implementations of public policy that bolster our collective good. We need to center the most vulnerable among us—the working poor and the disabled, those who live under racial and religious tyranny, the discriminated against and the oppressed. We need to house people and provide health, employment, creative arts and educational resources. We need to close the intersectional racial wealth gap. We are at another pivotal moment of reckoning about the immorality of our systems, and it's a good time to reimagine regulation, restoration and reparation from Big Tech too."[9] It's not direct reparations, but closing the tax code loopholes that allow tech companies to set up offshore accounts and avoid paying taxes would be another way of achieving financial compensation from Big Tech.

Thanks in part to muckraking journalists who are doing investigative deep dives into technology, there is a growing resistance to surveillance culture. The Markup, NBC News, ProPublica, and the *New York Times* are only some of the news organizations that are doing important work demystifying technology and exposing its many injustices. Kashmir Hill's 2019 Gizmodo story, "I Cut the 'Big Five' Tech Giants from My Life. It Was Hell," powerfully illustrated how it is practically impossible to engage in modern life without consuming Big Tech products. The Markup's 2021 story "The Secret Bias Hidden in Mortgage-Approval Algorithms" showed how people of color are denied mortgages at significantly higher rates than white people. Adam Satariano and Mike Isaac showed in the *New York Times* how a $500 million contract between Facebook and the consulting firm Accenture allowed profits to soar for the social media giant while negatively impacting the emotional and mental health of the commercial content moderators who review toxic content.[10] Clearview AI was first outed as a dystopian surveillance company by Kashmir Hill in the *New York Times* in January 2020. Since then, the company has been criticized widely for scraping more than three billion images from Facebook, YouTube, Venmo, and other websites and using the information to train facial recognition software. The ACLU has filed suit against Clearview AI under the Illinois Biometric Information and Privacy Act, alleging that community members have reason to fear loss of privacy, anonymity, and security because of the company's actions.[11]

Public opinion is turning against surveillance culture. So is some art. Indigenous futurism, Afrofuturism, Africanfuturism, Desi futurism, Arab futurism, Asian futurism, South Asian futurism, and Chicanafuturism are all gaining visibility in visual arts, fashion, film, and literature. Many of these movements engage with alternate visions of the future that move beyond our current glitchy, biased situation.

In addition to stories of people being stuck in long, complicated situations of algorithmic harm, I'm starting to hear more stories of people triumphing over algorithmic wrongs. One such example comes from Sujin Kim, a student who at the end of 2021 was a senior at the University of Michigan. Kim was eager to follow up her undergraduate political science studies with a PhD. She was applying to fifteen schools on the East and West Coasts. Her applications were in, she was on top of the deadlines, and the only thing that remained was to take the GRE, the standardized test required by many graduate schools. She scheduled it carefully: the scores were supposed to be returned in ten to fifteen days, so she picked a test date in early November, well before her December 1 application deadlines. The GRE would be administered remotely because it was 2021 and the COVID-19 pandemic was still raging in Michigan. She reserved a study room on campus so she could have a quiet spot to take the test without distractions. She knew that the test would be administered using proctoring software called ProctorU, and she knew the software would be finicky.

ProctorU, like ExamSoft or Proctorio, is one of the many AI-based remote proctoring software systems on the market. Unlike others, ProctorU has backed off from trying to use AI without human intervention. "We believe that only a human can best determine whether test-taker behavior is suspicious or violates test rules," said Scott McFarland, CEO of ProctorU, in a May 2021 press release. "Depending exclusively on AI and outside review can lead to mistakes or incorrect conclusions as well as create other problems."[12]

When the day came to take the test, everything seemed ordinary at first. She checked in using ProctorU, held up her ID to the camera so the software could take a picture of her face and her ID. The software took control of her desktop and she made sure the camera could see the door behind her, as instructed, so the remote proctor could see if anyone came into the room.

Problems arose quickly. "I got through the first section and the connection dropped," Kim said. She restarted the program and checked in again with the photo verification and began to write the required essay portion of the test. The connection dropped again. She tried to log back in, this time without success. Eventually, she connected with a human proctor who seemed flustered. She seemed to feel bad that Kim couldn't take her exam. The proctor took over Kim's desktop and checked her in again, and this time it worked. Kim finished the exam and the program issued her a set of preliminary scores. The preliminary scores would be verified by ETS and she would get her final scores in ten to fifteen days.

Ten days later, the scores had not arrived. Kim went home to New Jersey for Thanksgiving. Her parents and friends asked how graduate school applications were going, and she admitted that her GRE scores hadn't arrived. Friends who had taken the test at the same time had received their scores in a week. "I thought, okay, the scores are probably going to come out after fifteen days, because of the holiday," Kim told me. "I thought, 'thank God I built in extra time.'" On the fifteenth day, nothing arrived. "I stayed up till midnight to see if they would come out and they didn't," she said. "I didn't get an email. When I went into my portal, it just said, 'scores unavailable.' I emailed ETS. I didn't get anything. Then, I spent four days on the phone or trying to get through to them on the phone, on hold for literally hours, trying to figure out what was happening because no one would tell me." She skipped class on the fourth day so she could stay on hold, studying for hours in the library with one earphone in, hoping to connect with an ETS agent.

On November 30, the day before her first graduate school application deadlines, Kim got up at 5 a.m. so she could contact the central ETS customer service line and, she hoped, get in before the daily rush. She was on hold for five hours before someone picked up. "I got through to someone and they kept passing me around offices, like they were not clear about who deals with what," she said. Eventually, she connected with an agent who said there was a security hold on her scores. "She was working at home; I could hear her kid crying in the background," Kim said. "I asked her when I could expect the scores by. She said it depends on what the hold is for. I said, 'Okay, can I have documentation or something to send to the schools so I can tell them I'm not lying about my scores being delayed?' She said, 'It looks like there was a problem with your pictures, your ID photo, so that's usually two to eight weeks.' She said not to worry, the schools would give me extensions. She told me to just call the schools I was applying to and ask. Then she hung up."

Kim blamed herself. Maybe she had closed her eyes or been out of the frame in one of the verification photos? Could that be the reason for the security hold? She went on Reddit and the ProctorU website to look up security holds. "I assumed it was a person doing the verification," she said. "It's actually facial recognition. They use a biometric facial ID, and if the score is above a certain threshold match you can proceed. That's how I found out it was a tech problem."

Kim's story could have ended there—unable to get final scores, contacting fifteen different graduate programs to try to get extensions on her application deadlines. But she was not easily cowed. She had heard the story of Robert Williams, a fellow Michigan resident who had been wrongly identified as a criminal by a facial recognition program. And she had paid attention in school. Kim worked as a research assistant for Shobita Parthasarathy and Molly Kleinman at the University of Michigan's Ford Science, Technology, and Public Policy program. Parthasarathy, the program director, is a professor of public policy; Kleinman is the program's managing director. The lab had published a study the year before called "Cameras in the Classroom," which criticized facial recognition in education and called for a ban on its use. "Not only is the technology not suited to security purposes, but it also creates a web of serious problems beyond racial discrimination, including normalizing surveillance and eroding privacy, institutionalizing inaccuracy and creating false data on school life, commodifying data and marginalizing nonconforming students," Parthasarathy wrote about facial recognition.[13]

Kim told Kleinman about the situation. Kleinman tweeted about it. "What an incredibly stressful, time-wasting thing to impose on a young student, in the middle of a stressful semester and an ongoing pandemic," she wrote. "She, and other students like her, has to bear all of the risk of any failures of the system, meanwhile nobody takes responsibility. This outcome was entirely predictable. In fact, we (@STPP_UM) predicted it. By virtue of being young, a woman, and not white, my RA is exactly who we would expect to be hurt by the use of [facial recognition] and other automated systems in education."

The thread went viral in ethical AI circles. People called the situation Kafkaesque, Orwellian. The professoriate was outraged. "This student understood exactly what was happening and why it was wrong because she happens to work with a research team that literally wrote the report on why facial recognition in education should be banned," wrote Kleinman. "Imagine all the students who don't have that knowledge or access."

ETS seemed to take notice. Mysteriously, just before the December 1 deadline and after Kleinman's thread, ETS resolved its security hold and released Kim's final scores. Kim was relieved.

ProctorU's founder, Jarrod Morgan, entered the chat eight days later. "Hey Molly—Something isn't right here, as that's not how our process works," he tweeted. "@ShobitaP, I'm so sorry your scores haven't been released yet. If you can DM me some details, I'll get our compliance team working on this right away." Morgan seemed to be under the impression that Shobita Parthasarathy, a full professor and noted scholar of science and technology studies, was the one who was waiting on GRE scores.

When I spoke with Kim about the facial recognition fail that could have wrecked her chances of getting into graduate school, she was keenly aware of how institutional privilege and educational capital had helped her overcome this case of algorithmic bias. "I find it hard to believe it's just a coincidence. I'm sure the scores coming out had a lot to do with Molly advocating for me," Kim told me. "A lot of circumstances came together in a way that was very favorable for me in this situation. I'm sure this kind of problem happens to people all the time, and most people don't have that kind of recourse available to them. This could have gone so poorly if other people hadn't decided to be understanding. My professors went all in for me. Most people probably don't have that kind of fallback."

We debated what could have happened inside the software. She thought the problem was a check-in photo that had her face out of frame. I thought the problem was racism in the facial recognition software. In a NIST study, facial recognition technology was 10 to 100 times more likely to inaccurately identify a Black or East Asian face, compared to a white face. There is also a case in New Zealand where a man of Asian descent was unable to get his passport photo automatically approved because an AI repeatedly registered his eyes as being closed.[14] Both explanations were plausible.

People like to have a "good reason" for a decision, and algorithms rarely give one. I was explaining to a friend, a Spanish-language instructor, why A+-student Isabel Castañeda had been down-ranked by the IB grading algorithm, and she asked me why it happened. I explained the same thing I wrote earlier: the inputs to the algorithm, historical inequity, centuries of racist education in the United States, and so forth. "Yes, but *why?*" my friend asked. "If her teacher put in a good evaluation for her, wouldn't that be enough to make the algorithm assign her a good grade if the teacher's evaluation carries the most weight?" I started to say that algorithmic weighting isn't quite the same as the weights we assign in a syllabus, where we might say that the final project is worth 40 percent of the grade, homework is worth 50 percent, and class participation is worth 10 percent. Then, I stopped myself—because the algorithmic grading scheme actually *is* like the weights we assign in a syllabus, except it's more mathematically complicated. My friend's question was getting at something essential and frustrating about algorithmic explanations. There is rarely a reason for algorithmic decisions that will make sense to a human being. Unless a data scientist is willing to sit down and explain every feature of an algorithmic system, and provide the code and the training data and the social context in which the system was developed and deployed, it's hard and deeply unsatisfying to explain what happened. When you examine or audit a system, you get familiar enough with its inputs and outputs that it feels like you understand its reasoning. Without that understanding of the code and the training data and the social context, it feels opaque—like a black box. It takes an investment of time in order to understand what's going on, and the person asking for the reason also needs to invest in understanding all the dimensions of the system. Neither Isabel Castañeda nor Sujin Kim had all the information they would have needed to understand the complete dimensions of what happened—but they did know enough to know that something wasn't right. I find it hopeful that students can self-advocate in this regard. I hate that dealing with algorithmic failures is effectively a tax on anyone who isn't an able-bodied cisgender white male, but I do take it as a good sign that people are increasingly willing to call out algorithmic bias when it happens.

I thought more about what happened with Sujin Kim's test scores. The system failure seemed clear, the company worked quickly to address the problem once public pressure was applied, and Kim's self-advocacy paid off—but there was clearly a problem with the software that was much more than a glitch.

I was reminded of my intro to computer science professor who was trying to explain recursion, a concept that I was having trouble grasping conceptually. I kept asking questions, and he got more and more frustrated that I wasn't understanding. "Just pretend it's magic," he said. "If you don't understand what is happening, just pretend it's magic and write the code—it will work." I was embarrassed in the moment, and a little hurt, because I could tell he was annoyed that I didn't understand. But I listened to him anyway, and I wrote the code the way he suggested, and it worked, and eventually I figured out recursion. (It's nifty.) I have complicated feelings about that "magic" comment. I realized that I wanted to tell my friend to pretend the grading algorithm was magic, so I could stop explaining in that moment. But I also wanted to push through and explain better, because I didn't want my friend to feel abandoned intellectually the way I did when the professor got frustrated with me.

The thing about the magic is that it allows you to stop feeling stressed about understanding something in the moment. The key is to not stop there—you have to keep wondering, keep striving to understand. If you just stop with the magic, and trust that it works, you stop looking for flaws. So, magic works, until it doesn't. I recommend trusting that the reason for an algorithmic decision is some kind of preexisting social bias manifesting in subtle (or obvious) ways. It feels unsatisfying to lack a reason, and that dissatisfaction can fuel a drive to achieve social justice.

Tech is racist and sexist and ableist because the world is so. Computers just reflect the existing reality and suggest that things will stay the same—they predict the status quo. By adopting a more critical view of technology, and by being choosier about the tech we allow into our lives and our society, we can employ technology to stop reproducing the world as it is, and get us closer to a world that is truly more just.

Acknowledgments

I am grateful to so many people who contributed to this project. Thank you to the many folks who generously offered interview time and expertise over the past two years, including Richard Dahan, Uché Blackstock, Isabel Castañeda, Chris Gilliard, Kryztof Geras, Cori Zarek, Hana Schank, Julia Angwin, Albert Fox Cahn, Sujin Kim, Molly Kleinman, and Rolisa Tutwyler. To my husband and son: I love you so much. Thank you for letting me disappear to work on the book, and for your patience in listening to computational horror stories at the dinner table. To Elena Lahr-Vivaz: thank you for being my first and best reader. This book would never have come into being without the Writing with Randos writing group, who gather daily on Zoom at an alarmingly early hour. I especially thank Elizabeth Popp Berman, LaDale Winling, Christina Ho, Elly Truitt, Danna Agmon, Dan Hirschmann, Louise Seamster, Deana Rohlinger, Tanya Schlam, and Jordan Ellenberg. To my incredible agent, Tanya McKinnon: you are a marvel and I am delighted to be in your orbit. Thank you to my editor Gita Manaktala, Jessica Pellien, Deborah Cantor-Adams, Suraiya Jetha, and all the MIT Press folks. To Cathy O'Neil, Joy Buolamwini, Safiya Noble, and Ruha Benjamin: your work continues to inspire me and I value your friendship. I'm also grateful to the remarkable scholars at the Center for Critical Race & Digital Studies. Thank you to Solon Barocas, Avriel Epps-Darling, Sam Lavigne, Jacob Appel, Tom Adams, and Betty O'Neil for reading and contributing.

To Clay, Erika, Miles, Lawrence, Michaela, and Sue: thank you for caring for my family while I was in the hospital. To Janet, Kay, and Dean: thank you for your many years of companionship and support. To Mike, Beth, Warren, Candida, and Steven: thank you for the good times over truffles. To all the Alteveers, Kinseys, Hunts, and Moores: thanks for many years of friendship. To Delma, Ray, Lauren, Scott, Becca, BC, Andrea, and Ness: love you always, thanks for putting up with all of the links I sent to the group chat. I hope you note the number of items in the bibliography and admire my restraint in not sending all of them. Big hugs to Mary, Jeff, Christine, Katie, Will, Sophia, Liam, Stacie, Anthony, AJ, Mia, Miriam, Samira, Amelia, Rob, Kira, John, Jules, Jane, Ivan, Oliver, Poppy, and Marley. To Nancy, Evan, Ben, Nate, Clare, Rich, Theo, Ray, Ellie, Josh, Mattie, Sasha, Meeghan, David, Alex, and Ting: thank you for keeping me sane through the lockdowns and for always being willing to seek out a new swimming hole. To my knitting group and the WOC group chat: our talks and our sisterhood were a source of joy in the pandemic. To Sultan and Victor: thanks for helping me put my body back together. To my research assistants Isaac Robinson, Joshua Arrayales, and Huanjia Zhang: thank you, you were invaluable. I am grateful to my NYU colleagues, especially Perri Klass, Pamela Newkirk, Hilke Schellmann, Rachel Swarns, Meryl Gordon, Katie Roiphe, Charles Seife, Frankie Edozien, Yvonne Latty, Adam Penenberg, Dan Fagin, Jay Rosen, Eliza Griswold, Ivan Oransky, Charlton McIlwain, Melissa Lucas-Ludwig, Gianpaolo Baiocchi, Paula Chakkravarty, Faye Ginsburg, Ann Morning, and everyone in the NYU Alliance for Public Interest Technology. I wrote a lot of this book in a rural town where the power and wifi conk out at inconvenient times. I'm grateful to Hickory Hill Market, the only place in town with a generator, where I took a lot of Zoom meetings in the parking lot and finished editing this manuscript at a table in the deli.

In this book, I've mentioned or given small portraits of people I find inspirational. There are many I didn't name, who are doing work that inspires me, including Tressie McMillan Cottom, Dean Freelon, Chancey Fleet, Mary L. Gray, Harlo Holmes, Rashida Richardson, Lilly Irani, Xiaowei Wang, and Meredith Whitaker. In addition to *Weapons of Math Destruction*, *Coded Bias*, *Algorithms of Oppression*, *Race After Technology*, and *Artificial Unintelligence*, a variety of books, films, videos, podcasts, and artworks provide a deep dive into the myriad issues associated with algorithmic justice. In no particular order, I recommend:

Charlton McIlwain, *Black Software*

Kate Crawford, *Atlas of AI*

Mar Hicks, *Programmed Inequality*

Virginia Eubanks, *Automating Inequality*

Alberto Cairo, *How Charts Lie*

Sarah Brayne, *Predict and Surveil*

Dorothy Roberts, *Fatal Invention*

Simone Browne, *Dark Matters*

Visual art by Stephanie Dinkins, Mimi Onuoha, Nettrice Gaskins, or Amir Baradaran

I'm also grateful to you, the reader. Because you are reading this book, you are making the effort to understand how structural bias manifests in technology and to imagine how we can change it to make a difference in the world. Thank you. I hope you are going to take this knowledge and go out to make a better future.

Notes

Chapter 1

1. The IT folks are certain that it will make their jobs easier if you will please just use the technology that they signed a contract for, and they know it will work on your machine, and they have tech support for it, and they don't understand why you want to use this technology that is not supported by the central IT department. When you come to them to ask for troubleshooting help, they will ask—justifiably—if you have turned your computer off and then turned it on again.

2. The concept of a glitch as a site for illuminating social problems has been explored extensively in the academic field of science and technology studies. For the importance of the "glitch" concept in understanding how white supremacy operates inside the tech world, I am grateful to Ruha Benjamin, who explores the idea beautifully in her book *Race after Technology*. Benjamin's book inspired me to look deeper into several issues I started exploring in *Artificial Unintelligence*, and it helped me to gain a better understanding of both the racist soap dispenser and Shirley cards.

3. Noble, *Algorithms of Oppression*.

4. Benjamin, *Race after Technology*.

5. Hannah-Jones, Twitter post, July 7, 2021.

6. In *Algorithms of Oppression*, Safiya Noble called for a literature of Black feminist tech criticism. When I read that, I thought: "I would read the heck out of such a literature." Catherine Knight Steele's 2021 *Digital Black Feminism* answered the call, offering "a history of Black feminist technoculture in the United States and its ability to decenter white supremacy and patriarchy."

7. Nix, "Facebook Sees Slight Decline in Female Worker Representation."

Chapter 2

1. Ellenberg, *Shape*.

2. If we used a time scale, a million seconds is twelve days; a billion seconds is thirty-one years.

3. Ellenberg, again in *Shape*, has the best strategy on visualizing: "If you're finding it hard to imagine what a fourteen-dimensional landscape looks like, I recommend following the advice of Geoffrey Hinton, one of the founders of the modern theory of neural nets: 'Visualize a 3-space and say 'fourteen' to yourself very loudly. Everyone does it.'"

4. Hamilton, *Democracy's Detectives*.

5. Harney, "Large Numbers of Loan Applications Get Denied."

6. Aougab et al., "Math Boycotts Police"; Linder, "Why Hundreds of Mathematicians Are Boycotting Predictive Policing"; "Letter to AMS Notices."

7. Lorde, *Sister Outsider*.

8. Martinez and Kirchner, "The Secret Bias Hidden in Mortgage-Approval Algorithms."

Chapter 3

1. Williams, "I Was Wrongfully Arrested Because of Facial Recognition."

2. Jackson, "Challenging Facial Recognition Software."

3. Urban et al., "A Critical Summary of Detroit's Project Green Light."

4. Buolamwini, "Safe Face Pledge Launched."

5. Hill, "Wrongfully Accused by an Algorithm."

6. Detroit Police Department, "Facial Recognition Directive 307.5."

7. NASP, "The Shoplifting Problem."

8. Mack, "How Technology and Retail Trends Have Changed Shoplifting."

9. City of Detroit, "Project Green Light Detroit."

10. Fung and Metz, "This May Be America's First Known Wrongful Arrest Involving Facial Recognition."

11. Williams, "I Was Wrongfully Arrested."

12. Kendi, "'The Difference between Being 'Not Racist' and Antiracist.'"

13. Barbican Centre, "Joy Buolamwini."

14. Buolamwini, "How I'm Fighting Bias in Algorithms."

15. Zuckerman, "To the Future Occupants of My Office."

16. Singer and Metz, "Many Facial-Recognition Systems Are Biased."

17. Goodwin, Testimony before the Subcommittee on Crime, Terrorism, and Homeland Security.

18. Goodwin, Testimony before the Subcommittee on Crime, Terrorism, and Homeland Security.

19. US Government Accountability Office, "Facial Recognition Technology."

20. Warzel and Thompson, "They Stormed the Capitol"; Treisman, "Bumble Blunder."

21. Harwell, "Wrongfully Arrested Man Sues Detroit Police."

22. Ng, "Police Say They Can Use Facial Recognition."

23. Anderson, "Controversial Detroit Facial Recognition Got Him Arrested."

24. Hill, "Another Arrest, and Jail Time."

Chapter 4

1. Stroud, "An Automated Policing Program Got This Man Shot Twice."

2. "Predictive Policing"; Heaven, "Predictive Policing Algorithms Are Racist."

3. Baltimore used a variety of CompStat software called Citistat. Originally developed to reduce employee absenteeism, Citistat launched in June 2000 and was deployed on the Bureau of Solid Waste within Public Works. It eventually spread to govern other City of Baltimore departments, including the police. Though Citistat was based on NYPD's CompStat, it was relatively low-tech: as of 2007, it relied on data collected in Microsoft Excel, presentations presented in PowerPoint, and maps generated with ESRI's ArcView software, according to a report from the Center for American Progress. While data-driven practices are undeniably useful to save taxpayer dollars in government and to help the authorities be good stewards of public funds, problems arise when people start imagining that data-driven practices are appropriate to solve or predict *every* problem.

4. Brayne, *Predict and Surveil*.

5. Mazmanian and Beckman, "'Making' Your Numbers."

6. McIlwain, *Black Software*.

7. McGrory and Bedi, "Targeted"; Bedi and McGrory, "Pasco's Sheriff Uses Grades and Abuse Histories"; McGrory and Bedi, "The Man behind the Machine"; Pulitzer Center for Health Journalism, "Pulitzer Spotlight."

8. Pulitzer Center for Health Journalism, "Pulitzer Spotlight."

9. Brayne, *Predict and Surveil*.

10. Longley, "The History of Modern Policing."

11. American Civil Liberties Union, "Why Is It so Hard to Hold Police Accountable?"

12. Browne, *Dark Matters*.

13. Scannell, "Broken Windows, Broken Code."

14. Ray, "How Can We Enhance Police Accountability?"

15. Angwin et al., "Machine Bias."

16. Kleinberg, Mullainathan, and Raghavan, "Inherent Trade-Offs in the Fair Determination of Risk Scores."

17. Ensign et al., "Runaway Feedback Loops."

18. Richardson, "Racial Segregation and the Data-Driven Society."

19. Petty, Twitter post, June 2, 2021.

20. Clifton, Lavigne, and Tseng, "White Collar Crime Risk Zones."

21. Clifton, Lavigne, and Tseng, "White Collar Crime Risk Zones."

22. Eisinger and Kiel, "The Top 0.5% Underpay $50 Billion a Year in Taxes."

23. Eisinger, Ernsthausen, and Kiel, "The Secret IRS Files."

24. Ray, "How Can We Enhance Police Accountability?"

25. Melamed, "A Leaked Memo Suggests Philly Police Use Vehicle Stops"; New York Civil Liberties Union, "Stop-and-Frisk Data."

26. Ryan-Mosley and Strong, "The Activist Dismantling Racist Police Algorithms."

27. Farivar, "Cops Decide to Collect Less License Plate Data."

28. Farivar, Twitter post, May 27, 2021; Sciacca, "Oakland Police Give FBI 'Unfettered Access.'"

29. Crawford, *Atlas of AI*; Lécuyer, "From Clean Rooms to Dirty Water"; Burrington, "Light Industry."

30. Rex, "N.Y.P.D. Will Stop Using Robot Dog."

31. Chandran, "Afghans Scramble to Delete Digital History."

32. Clayton, "How Eugenics Shaped Statistics."

33. Zuberi, *Thicker Than Blood*.

34. IBM, "AI Governance."

35. Amrute, "Of Techno-Ethics and Techno-Affects."

36. O'Neil, *Weapons of Math Destruction.*

37. Clayton, "How Eugenics Shaped Statistics."

Chapter 5

1. Caspary et al., "International Baccalaureate National Trends."

2. Lough, "IB Results"; Zaman, "IB to Make Adjustments ."

3. Lecher and Varner, "Remote Learning during the Pandemic Has Hit Vulnerable Students the Hardest."

4. New York City Department of Education, "School Quality Reports: Using 'Comparison Group' Results to Better Understand a School's Performance."

5. . Math and stats people are often okay with everything evening out in the aggregate, but they are also often people with a great deal of privilege. The average salary for a government mathematician in 2019 was over $140,000, which is almost three times the average US worker's salary.

6. Hess, "Rich Students Get Better SAT Scores."

7. Ward and Moran, "Thousands of California Bar Exam Takers"; Spiezio, "Michigan Software Crash"; Sloan, "Plaintiffs' Firm Eyes Class Action ."

8. Shapiro, "When the World Went Remote."

9. Tate, "Why Aren't Schools Using the Apps They Pay For?"

Chapter 6

1. Apple Newsroom, "Apple Brings Everyone Can Code to Schools Serving Blind and Deaf Students."

2. This was the price for a 140-watt USB-C power adapter in October 2021, which I tried to buy because the battery in my MacBook Pro stopped taking a charge four days before I was scheduled to turn in the manuscript for this book. The adapter was back-ordered and would take a month to arrive.

3. Data Visualization Society, *Accessibility Fireside Chat.*

4. US General Services Administration, "Universal Design: What Is It?"

5. Girma, *Haben.*

6. Data Visualization Society, *Accessibility Fireside Chat*.

7. Quinn et al., "Human Rights and Disability."

8. US General Services Administration, "Universal Design: What Is It?"

9. Edwards, TikTok post, June 6, 2021.

10. Roy, "When We Design for Disability, We All Benefit."

11. Benjamin, *Race after Technology*.

12. Costanza-Chock, *Design Justice*.

13. Gibson and Williams, "Who's in Charge?"

14. "Color Blindness."

15. Cravit, "How to Use Color Blind Friendly Palettes"; "What Is Color Blindness?"

16. Data Visualization Society, *Accessibility Fireside Chat*.

17. Data Visualization Society, *Accessibility Fireside Chat*.

18. Girma, *Haben*.

19. Foreman, "We Can't Automate Alt Text."

20. Challenge Solutions, "A Comparison of Three Screen Readers."

21. Girma, *Haben*.

22. In my interviews for this book, I asked interviewees how they preferred to be identified, especially regarding capitalization and pronouns. When they didn't have a stated preference, I went with the self-description in their published works or in their social media profiles. I chose to capitalize Blind and Deaf throughout the manuscript, just as I chose to capitalize Black, because naming a group and capitalizing it suggests empowerment to me at this time. Schools of thought around capitalization and identity have changed over the course of my lifetime, and I hope to have the opportunity to update my own style conventions as social conventions change in the future.

23. Raji, Scheuerman, and Amironesei, "You Can't Sit with Us."

24. Jackson, "A Community Response to a #DisabilityDongle."

25. "Disabled People Want Disability Design."

26. Young, "The Least Livable Body in America's Most Livable City."

27. Imani, "The Call Is Coming from Inside the House."

28. Whittaker et al., "Disability, Bias, and AI."

29. Imani, "The Call Is Coming from Inside the House."

Chapter 7

1. Hicks, "Hacking the Cis-Tem."

2. Bowker and Star, *Sorting Things Out.*

3. D'Ignazio and Klein, "What Gets Counted Counts."

4. Bivens, "The Gender Binary Will Not Be Deprogrammed."

5. Hurtle, "I'm a Trans Woman."

6. Burrington, Twitter post, September 18, 2019.

7. Johnson, "The Role of the Digital Computer."

8. Stevens, Hoffmann, and Florini, "The Unremarked Optimum."

9. Harmon, "Which Box Do You Check?"

10. Keyes, "The Misgendering Machines."

11. NYU Communications, "Gender Identity in Albert."

12. Hicks, "Hacking the Cis-Tem."

13. Hill, "NY Social Service Agency Sued."

14. Hicks, "Hacking the Cis-Tem."

15. Waldron and Medina, "TSA's Body Scanners Are Gender Binary."

16. Costanza-Chock, *Design Justice.*

Chapter 8

1. Richardson, "Patterns and Trends in Age-Specific Black-White Differences."

2. Goodnough, "Finding Good Pain Treatment Is Hard."

3. Salam, "For Serena Williams, Childbirth Was a Harrowing Ordeal."

4. McDonald, "'Believe Me' Means Believing That Black Women Are People."

5. Skloot, *The Immortal Life of Henrietta Lacks.*

6. Sjoding et al., "Racial Bias in Pulse Oximetry Measurement"; Goold, Sjoding, and Valley, "Black People Are Three Times More Likely to Experience Pulse Oximeter Errors"; Villarosa, *Under the Skin: Racism, Inequality, and the Health of a Nation.*

7. Young, "Racism Makes Me Question Everything."

8. Bui and Liu, "Using AI to Help Find Answers."

9. Bui and Liu, "Using AI to Help Find Answers."

10. Liu et al., "A Deep Learning System."

11. Feathers, "Google's New Dermatology App."

12. Noble, *Algorithms of Oppression.*

13. Agénor et al., "Developing a Database of Structural Racism–Related State Laws for Health Equity Research and Practice in the United States"; Christensen, Manley, and Resendez, "Medical Algorithms Are Failing Communities of Color"; Thomas, "Medicine's Machine Learning Problem"; Strickland, "Racial Bias Found in Algorithms"; Tang, "A New Study Found Many Clinical Algorithms Are Still Subjected to Racial Biases"; Kolata, "Many Medical Decision Tools Disadvantage Black Patients"; Braun et al., "Racial Categories in Medical Practice."

14. National Kidney Foundation, "Understanding African American and Non-African American eGFR Laboratory Results."

15. Grubbs, "Precision in GFR Reporting."

16. Heffron et al., "Trainee Perspectives on Race."

17. National Kidney Foundation, "Establishing a Task Force"; National Kidney Foundation, "Race and EGFR."

18. Heffron et al., "Trainee Perspectives on Race."

19. Roberts, Twitter post, April 30, 2021.

20. Mukherjee, "A.I. Versus M.D."

21. Frankle et al., "Values in Science and Engineering of ML Research."

22. Bergstrom and West, *Calling Bullshit.*

23. "U of T Neural Networks Start-Up Acquired by Google."

24. Kamara, "Crypto for the People."

25. Ferryman and Pitcan, "Fairness in Precision Medicine."

26. Kadambi, "Achieving Fairness in Medical Devices."

27. Wallis, "Fixing Medical Devices That Are Biased against Race or Gender."

28. Gopal et al., "Implicit Bias in Healthcare."

29. Roberts, "Race Correction."

Chapter 9

1. Robbins and Brodwin, "Patients Aren't Being Told about the AI Systems."

2. Geras et al., "High-Resolution Breast Cancer Screening."

3. Wu et al., "The NYU Breast Cancer Screening Dataset v1.0."

4. Crawford, *Atlas of AI.*

5. Geras et al., "High-Resolution Breast Cancer Screening."

6. Isaac Robinson offered an excellent explanation of the situation:

> There are many rulers we use to measure the effectiveness and efficacy of machine learning algorithms, what we call error metrics. At times the simplest and most complex is one called AUC, which measures the area under the ROC curve plotting true positive rates versus true negative rates.
>
> That can be confusing, so let's break down what the positives and negatives mean. In statistics we have a set of terms which are permutations of these four words:
>
> false/true + positive/negative
>
> Let's start with the second part. Most of what machine learning algorithms do is tell you yes or no. Yes this is a cat, no you should not give that person a loan, yes that user does want to buy new face cream. Positive or negative are really just talking about when the algorithm is saying yes, and when the algorithm is saying no. In our case, a positive result is any one where the algorithm thinks there is cancer present in the image.
>
> Now, false and true modify positive and negative. Whereas positive and negative are concerned with what the model is saying, false and true are more concerned with what the actual truth is. False means that the algorithm predicted wrong, and true means the algorithm predicted right. So a false positive means the algorithm said there was cancer, but ground truth says there was not. A false negative means the algorithm said there was no cancer, but there was.
>
> If we plot false positive against true positive rates, we get what we call the ROC curve. This basically represents a tradeoff of how many false positives are we willing to give in order to assure our true positive rate is above a certain threshold. Measuring the area of this curve gives us one possible metric for quantifying how good our algorithm is. For those who are interested, this measurement can be roughly thought of as the probability that the model ranks a random positive example more highly than a random negative example.
>
> As you can imagine, this tradeoff becomes an essential consideration for algorithms that make decisions about people's lives. How many people are we willing to scare by telling them they have cancer in order to assure we never tell someone who does have it that they don't? Perhaps an even bigger questions—who gets to make that decision?

7. Wu et al., "How Medical AI Devices Are Evaluated."

8. Wu et al., "How Medical AI Devices Are Evaluated."

9. Lebovitz, Lifshitz-Assaf, and Levina, "To Engage or Not to Engage with AI."

10. Black and Richmond, "Improving Early Detection of Breast Cancer"; Birhane, "Algorithmic Colonization of Africa."

11. Galukande and Kiguli-Malwadde, "Rethinking Breast Cancer Screening Strategies"; Sankaranarayanan, "Screening for Cancer in Low- and Middle-Income Countries"; Ross, "Machine Learning Is Booming in Medicine."

Chapter 10

1. McGuinness and Schank, *Power to the Public.*

2. US Digital Service, "USDS Alumni Network."

3. Raji, Twitter post, June 28, 2021.

4. Jillson, "Aiming for Truth, Fairness, and Equity."

5. Verma and Rubin, "Fairness Definitions Explained."

6. Kleinberg, Mullainathan, and Raghavan, "Inherent Trade-Offs in the Fair Determination of Risk Scores."

7. Buolamwini and Gebru, "Gender Shades."

8. There's a whole field called Explainable AI, which has largely . . . not made AI explainable. Most of the explanations from Explainable AI are highly technical, meaning they are comprehensible to a small community rather than to the population at large.

9. Benjamin, *Race after Technology.*

10. Metz and Hill, "Here's a Way to Learn If Facial Recognition Systems Used Your Photos."

11. Quoted in Martinez and Carollo, "Markup Investigation Cited by Officials."

12. Ng, "Senator Calls Facebook Response on Discriminatory Ads 'Inadequate.'"

13. Brooks, "Disparity in Home Lending Costs Minorities Millions."

14. *Wall Street Journal*, "The Facebook Files."

15. Zakrzewski et al., "How Facebook Neglected the Rest of the World, Fueling Hate Speech and Violence in India."

16. AI Fairness 360, "Home."

17. Wilson et al., "Building and Auditing Fair Algorithms."

18. Cahn, "The First Effort to Regulate AI Was a Spectacular Failure."

19. Ryan-Mosley and Strong, "The Activist Dismantling Racist Police Algorithms."

Chapter 11

1. Belson, "Changes to N.F.L. Settlement Could Aid Thousands."

2. "Wyden, Booker and Clarke Introduce Algorithmic Accountability Act of 2022."

3. Jillson, "Aiming for Truth, Fairness, and Equity in Your Company's Use of AI."

4. Green, "How Labor Supports Union Organizing."

5. Jackson, Bailey, and Foucault Welles, *#HashtagActivism*.

6. "#AppleToo"; Schiffer, "Apple Employees Are Organizing."

7. Rhinesmith, "Joint Civil Society Statement."

8. Achiume et al., "Racial Discrimination and Emerging Digital Technologies."

9. Noble, "The Loss of Public Goods to Big Tech."

10. Roberts, *Behind the Screen.*

11. American Civil Liberties Union, "ACLU v. Clearview AI."

12. "ProctorU to Discontinue Exam Integrity Services."

13. "Cameras in the Classroom."

14. Bushwick, "How NIST Tested Facial Recognition Algorithms for Racial Bias"; Griffiths, "New Zealand Passport Robot Thinks This Asian Man's Eyes Are Closed"; "Passport Facial Recognition Checks Fail to Work with Dark Skin"; Dunne, "Man Stunned as Passport Photo Check Sees Lips as Open Mouth."

Bibliography

Achiume, Tendayi, and UN Human Rights Council. "Racial Discrimination and Emerging Digital Technologies: A Human Rights Analysis," June 18, 2020. Accessed July 20, 2022. https://digitallibrary.un.org/record/3879751.

Agénor, Madina, Carly Perkins, Catherine Stamoulis, Rahsaan D. Hall, Mihail Samnaliev, Stephanie Berland, and S. Bryn Austin. "Developing a Database of Structural Racism–Related State Laws for Health Equity Research and Practice in the United States." *Public Health Reports* 136, no. 4 (February 22, 2021). https://doi.org/10.1177 /0033354920984168.

AI Fairness 360. "Home." Accessed October 1, 2021. https://ai-fairness-360.org.

American Civil Liberties Union. "ACLU v. Clearview AI: Order Denying Motion to Dismiss." Accessed August 31, 2021. https://www.aclu.org/legal-document/aclu-v -clearview-ai-order-denying-motion-dismiss.

American Civil Liberties Union. "Why Is It so Hard to Hold Police Accountable?" At Liberty podcast, June 3, 2020. Accessed May 24, 2021. https://www.aclu.org/podcast /why-it-so-hard-hold-police-accountable-ep-102.

Amrute, Sareeta. "Of Techno-Ethics and Techno-Affects." *Feminist Review* 123, no. 1 (November 2019): 56–73. https://doi.org/10.1177/0141778919879744.

Anderson, Elisha. "Controversial Detroit Facial Recognition Got Him Arrested for a Crime He Didn't Commit." *Detroit Free Press*, July 10, 2020. https://www.freep.com /story/news/local/michigan/detroit/2020/07/10/facial-recognition-detroit-michael -oliver-robert-williams/5392166002.

Angwin, Julia, Jeff Larson, Surya Mattu, and Lauren Kirchner. "Machine Bias." *ProPublica*, May 23, 2016. https://www.propublica.org/article/machine-bias-risk -assessments-in-criminal-sentencing.

Aougab, Tarik, Federico Ardila, Jayadev Athreya, Edray Goins, Christopher Hoffman, Autumn Kent, Lily Khadjavi, Cathy O'Neil, Priyam Patel, and Katrin Wehrheim. "Math Boycotts Police." June 15, 2020. https://www.math-boycotts-police.net.

Apple Newsroom. "Apple Brings Everyone Can Code to Schools Serving Blind and Deaf Students." Accessed October 29, 2021. https://www.apple.com/newsroom /2018/05/apple-brings-everyone-can-code-to-schools-serving-blind-and-deaf -students.

"#AppleToo," August 2021. https://appletoo.us.

Barbican Centre. "Joy Buolamwini: Examining Racial and Gender Bias in Facial Analy-sis Software." Google Arts & Culture. Accessed January 13, 2022. https://artsandculture .google.com/exhibit/joy-buolamwini-examining-racial-and-gender-bias-in-facial -analysis-software/LgKCaNKAVWQPJg.

Bedi, Neil, and Kathleen McGrory. "Pasco's Sheriff Uses Grades and Abuse Histories to Secretly Label Kids Potential Criminals." *Tampa Bay Times*, November 19, 2020. Accessed July 21, 2021. https://projects.tampabay.com/projects/2020/investigations /police-pasco-sheriff-targeted/school-data.

Belson, Ken. "Changes to N.F.L. Settlement Could Aid Thousands of Black Players' Claims." *New York Times*, March 4, 2022. https://www.nytimes.com/2022/03/04/sports /football/nfl-race-norming-concussions-settlement.html.

Benjamin, Ruha. *Race after Technology: Abolitionist Tools for the New Jim Code*. Med-ford, MA: Polity, 2019.

Bergstrom, Carl T., and Jevin D. West. *Calling Bullshit: The Art of Skepticism in a Data-Driven World*. New York: Random House, 2020.

Birhane, Abeba. "Algorithmic Colonization of Africa." *SCRIPT-Ed* 17, no. 2 (August 6, 2020): 389–409. https://doi.org/10.2966/scrip.170220.389.

Bivens, Rena. "The Gender Binary Will Not Be Deprogrammed: Ten Years of Coding Gender on Facebook." SSRN Scholarly Paper. Rochester, NY: Social Science Research Network, December 27, 2015. https://doi.org/10.2139/ssrn.2431443.

Black, Eleanor, and Robyn Richmond. "Improving Early Detection of Breast Cancer in Sub-Saharan Africa: Why Mammography May Not Be the Way Forward." *Globalization and Health* 15, no. 1 (January 8, 2019): 3. https://doi.org/10.1186/s12992-018-0446-6.

Bowker, Geoffrey C., and Susan Leigh Star. *Sorting Things Out: Classification and Its Consequences*. Cambridge, MA: MIT Press, 1999.

Braun, Lundy, Anne Fausto-Sterling, Duana Fullwiley, Evelynn M. Hammonds, Alondra Nelson, William Quivers, Susan M. Reverby, and Alexandra E. Shields. "Racial Categories in Medical Practice: How Useful Are They?" *PLoS Medicine* 4, no. 9 (September 25, 2007): e271. https://doi.org/10.1371/journal.pmed.0040271.

Brayne, Sarah. *Predict and Surveil: Data, Discretion, and the Future of Policing.* New York: Oxford University Press, 2021.

Brooks, Kristopher J. "Disparity in Home Lending Costs Minorities Millions, Researchers Find." *CBS News Moneywatch*, November 15, 2019. https://www.cbsnews .com/news/mortgage-discrimination-black-and-latino-paying-millions-more-in -interest-study-shows.

Browne, Simone. *Dark Matters: On the Surveillance of Blackness.* Durham, NC: Duke University Press, 2015. https://doi.org/10.1215/9780822375302.

Bui, Peggy, and Yuan Liu. "Using AI to Help Find Answers to Common Skin Conditions." Google Keyword: Health, May 18, 2021. Accessed September 30, 2021. https://blog.google/technology/health/ai-dermatology-preview-io-2021.

Buolamwini, Joy. "How I'm Fighting Bias in Algorithms." TedXBeaconStreet, November 2016. Accessed October 24, 2021. https://www.ted.com/talks/joy_buolamwini _how_i_m_fighting_bias_in_algorithms/transcript.

Buolamwini, Joy. "Safe Face Pledge Launched to Prevent Abuse and Lethal Use of Facial Recognition Technology." *Medium* (blog), December 13, 2018. https://medium.com/@ Joy.Buolamwini/safe-face-pledge-launched-to-prevent-abuse-and-lethal-use-of-facial -recognition-technology-31d408a713df.

Buolamwini, Joy, and Timnit Gebru. "Gender Shades: Intersectional Accuracy Disparities in Commercial Gender Classification." *Proceedings of Machine Learning Research* 81 (2018): 1–15.

Burrington, Ingrid. "Light Industry: Toxic Waste and Pastoral Capitalism." *E-Flux*, no. 74 (June 2016). https://www.e-flux.com/journal/74/59781/light-industry-toxic -waste-and-pastoral-capitalism.

Burrington, Ingrid. Twitter post, September 18, 2019, 11:58 pm. Accessed October 2, 2021. https://twitter.com/lifewinning/status/1174533090548436994.

Bushwick, Sophie. "How NIST Tested Facial Recognition Algorithms for Racial Bias." *Scientific American*, December 27, 2019. Accessed February 27, 2022. https://www .scientificamerican.com/article/how-nist-tested-facial-recognition-algorithms-for -racial-bias.

Cahn, Albert Fox. "The First Effort to Regulate AI Was a Spectacular Failure." *Fast Company*, November 26, 2019. https://www.fastcompany.com/90436012/the-first -effort-to-regulate-ai-was-a-spectacular-failure.

"Cameras in the Classroom: Facial Recognition Technology in Schools." Webinar, Gerald R. Ford School of Public Policy, University of Michigan. Accessed February 27, 2022. https://fordschool.umich.edu/video/2020/cameras-classroom-facial -recognition-technology-schools-webinar.

Caspary, Kyra, Katrina Woodworth, Kaeli Keating, and Janelle Sands. "International Baccalaureate National Trends for Low-Income Students 2008–2014." Menlo Park, CA: SRI International, July 2015. https://www.ibo.org/globalassets/publications/ib -research/dp/ib-and-low-income-students-report-sri-en.pdf.

Challenge Solutions. "A Comparison of Three Screen Readers: JAWS, NVDA, and Voiceover." YouTube, September 21, 2020. Accessed August 11, 2021. https://www .youtube.com/watch?v=9_K5-4ngDtE.

Chandran, Rina. "Afghans Scramble to Delete Digital History, Evade Biometrics." Reuters, August 17, 2021. https://www.reuters.com/article/afghanistan-tech-conflict -idUSL8N2PO1FH.

Christensen, Donna M., Jim Manley, and Jason Resendez, "Medical Algorithms Are Failing Communities of Color." *Health Affairs*, September 9, 2021. Accessed March 3, 2022. https://www.healthaffairs.org/do/10.1377/forefront.20210903.976632 /full.

City of Detroit. "Project Green Light Detroit." Accessed April 29, 2021. https:// detroitmi.gov/departments/police-department/project-green-light-detroit.

Clayton, Aubrey. "How Eugenics Shaped Statistics." Nautilus, October 28, 2020. http://nautil.us/issue/92/frontiers/how-eugenics-shaped-statistics.

Clifton, Brian, Sam Lavigne, and Francis Tseng. "White Collar Crime Risk Zones." *The New Inquiry*, March 2017. Accessed July 22, 2020. https://whitecollar.thenewinquiry .com.

"Color Blindness." National Eye Institute. Accessed November 8, 2021. https:// www.nei.nih.gov/learn-about-eye-health/eye-conditions-and-diseases/color -blindness.

Costanza-Chock, Sasha. *Design Justice: Community-Led Practices to Build the Worlds We Need*. Cambridge, MA: MIT Press, 2020.

Cravit, Rachel. "How to Use Color Blind Friendly Palettes to Make Your Charts Accessible." *Venngage* (blog), August 21, 2019. https://venngage.com/blog/color-blind -friendly-palette.

Crawford, Kate. *Atlas of AI: Power, Politics, and the Planetary Costs of Artificial Intelligence*. New Haven, CT: Yale University Press, 2021.

Data Visualization Society. "Accessibility Fireside Chat." YouTube, July 28, 2021. Accessed July 29, 2021. https://www.youtube.com/watch?v=Aqx5PQwds80.

Detroit Police Department. "Facial Recognition Directive 307.5." July 25, 2019. https://detroitmi.gov/sites/detroitmi.localhost/files/2019-07/FACIAL%20RECOGNI-TION%20Directive%20307.5_0.pdf.

D'Ignazio, Catherine, and Lauren Klein. "What Gets Counted Counts." In *Data Feminism,* March 16, 2020. https://data-feminism.mitpress.mit.edu/pub/h1w0nbqp /release/3.

"Disabled People Want Disability Design—Not Disability Dongles." CBC Radio, November 8, 2019. https://www.cbc.ca/radio/spark/disabled-people-want-disability -design-not-disability-dongles-1.5353131.

Dunne, John. "Man Stunned as Passport Photo Check Sees Lips as Open Mouth." *Evening Standard*, September 19, 2019. https://www.standard.co.uk/news/uk/man -stunned-as-passport-photo-check-sees-lips-as-open-mouth-a4241456.html.

Edwards, Lucy. TikTok post, June 6, 2021. https://www.tiktok.com/@lucyedwards /video/6970764584269417733.

Eisinger, Jesse, Jeff Ernsthausen, and Paul Kiel. "The Secret IRS Files: Trove of Never-before-Seen Records Reveal How the Wealthiest Avoid Income Tax." *ProPublica,* June 8, 2021. https://www.propublica.org/article/the-secret-irs-files-trove-of-never-before-seen -records-reveal-how-the-wealthiest-avoid-income-tax.

Eisinger, Jesse, and Paul Kiel. "The Top 0.5% Underpay $50 Billion a Year in Taxes and Crushed the IRS Plan to Stop Them." *ProPublica*, April 5, 2019. https://www .propublica.org/article/ultrawealthy-taxes-irs-internal-revenue-service-global-high -wealth-audits?token=ernzrMMkK0V6INHywYhNmRrQnXa23xyN.

Ellenberg, Jordan. *Shape: The Hidden Geometry of Information, Biology, Strategy, Democracy, and Everything Else.* New York: Penguin Press, 2021.

Ensign, Danielle, Sorelle A. Friedler, Scott Neville, Carlos Scheidegger, and Suresh Venkatasubramanian. "Runaway Feedback Loops in Predictive Policing." *Proceedings of the 1st Conference on Fairness, Accountability and Transparency*, in *Proceedings of Machine Learning Research* 81 (2018):160–171. https://proceedings.mlr.press/v81 /ensign18a.html.

Farivar, Cyrus. "Cops Decide to Collect Less License Plate Data after 80GB Drive Got Full." Ars Technica, August 26, 2015. https://arstechnica.com/tech-policy/2015/08 /cops-decide-to-collect-less-license-plate-data-after-80gb-drive-got-full.

Farivar, Cyrus. Twitter post, May 27, 2021, 11:53 pm. https://twitter.com/cfarivar /status/1398125349691564041.

Feathers, Todd. "Google's New Dermatology App Wasn't Designed for People with Darker Skin." Vice, May 20, 2021. Accessed June 9, 2021. https://www.vice.com/en /article/m7evmy/googles-new-dermatology-app-wasnt-designed-for-people-with -darker-skin.

Ferryman, Kadija, and Mikaela Pitcan. "Fairness in Precision Medicine." Data & Society, February 2018, 54. https://datasociety.net/wp-content/uploads/2018/02 /DataSociety_Fairness_In_Precision_Medicine_Feb2018.pdf.

Foreman, Holden. "We Can't Automate Alt Text. Here Are Some Mistakes, Lessons and What We Can Do for Accessibility." Medium, July 22, 2021. https://washpost. engineering/we-cant-automate-alt-text-here-are-some-mistakes-lessons-and-what-we-can-do-for-accessibility-4e8631c2b3.

Frankle, Jonathan, Sandra Wachter, Shakir Mohamed, Emily Dinan, Danielle Belgrave, Meredith Broussard, and Silvia Chiappa. "Values in Science and Engineering of ML Research." Presented at the Science and Engineering of Deep Learning workshop, ICLR 2021, May 7, 2021. https://iclr.cc/virtual/2021/workshop/2128.

Fung, Brian, and Rachel Metz. "This May Be America's First Known Wrongful Arrest Involving Facial Recognition." CNN, June 24, 2020. https://www.cnn.com/2020/06 /24/tech/aclu-mistaken-facial-recognition/index.html.

Galukande, M., and E. Kiguli-Malwadde. "Rethinking Breast Cancer Screening Strategies in Resource-Limited Settings." *African Health Sciences* 10, no. 1 (March 2010): 89–92.

Geras, Krzysztof J., Stacey Wolfson, Yiqiu Shen, Nan Wu, S. Gene Kim, Eric Kim, Laura Heacock, Ujas Parikh, Linda Moy, and Kyunghyun Cho. "High-Resolution Breast Cancer Screening with Multi-View Deep Convolutional Neural Networks." June 27, 2018. http://arxiv.org/abs/1703.07047.

Gibson, Amelia, and Rua Williams. "Who's in Charge? Information Technology and Disability Justice in the United States." Just Tech, Social Science Research Council, March 1, 2022. https://doi.org/10.35650/JT.3030.d.2022.

Girma, Haben. *Haben: The Deafblind Woman Who Conquered Harvard Law.* New York: Twelve, 2019.

Goodnough, Abby. "Finding Good Pain Treatment Is Hard. If You're Not White, It's Even Harder." *New York Times*, August 10, 2016. https://www.nytimes.com/2016/08 /10/us/how-race-plays-a-role-in-patients-pain-treatment.html.

Goodwin, Gretta L. Testimony before the Subcommittee on Crime, Terrorism, and Homeland Security, Committee on the Judiciary, House of Representatives, Pub. L. No. GAO-21-105309 (2021). https://docs.house.gov/meetings/JU/JU08/20210713 /113906/HMTG-117-JU08-Wstate-GoodwinG-20210713.PDF.

Goold, Susan Dorr, Michael Sjoding, and Thomas Valley. "Black People Are Three Times More Likely to Experience Pulse Oximeter Errors." School of Public Health, University of Michigan. January 20, 2021. Accessed October 19, 2021. https://sph .umich.edu/pursuit/2021posts/black-people-are-three-times-more-likely-to-experience -pulse-oximeter-errors.html.

Gopal, Dipesh P, Ula Chetty, Patrick O'Donnell, Camille Gajria, and Jodie Blackadder-Weinstein. "Implicit Bias in Healthcare: Clinical Practice, Research and Decision Making." *Future Healthcare Journal* 8, no. 1 (March 2021): 40–48. https://doi.org/10.7861/fhj.2020-0233.

Green, Ken. "How Labor Supports Union Organizing at Big Tech Companies." *Union-Track* (blog), June 15, 2021. https://www.uniontrack.com/blog/big-tech-unions.

Griffiths, James. "New Zealand Passport Robot Thinks This Asian Man's Eyes Are Closed." CNN, December 8, 2016. https://www.cnn.com/2016/12/07/asia/new-zealand-passport-robot-asian-trnd/index.html.

Grubb, Vanessa. "Precision in GFR Reporting: Let's Stop Playing the Race Card." *Clinical Journal of the American Society of Nephrology* 15, no. 8 (August 2020): 1201–1202.

Hamilton, James. *Democracy's Detectives: The Economics of Investigative Journalism.* Cambridge, MA: Harvard University Press, 2016.

Hannah-Jones, Nikole. Twitter post, July 7, 2021, 11:14 am. https://twitter.com/nhannahjones/status/1412792267392290826.

Harmon, Amy. "Which Box Do You Check? Some States Are Offering a Nonbinary Option." *New York Times*, May 29, 2019. https://www.nytimes.com/2019/05/29/us/nonbinary-drivers-licenses.html.

Harney, Kenneth R. "Large Numbers of Loan Applications Get Denied. But for Blacks, Hispanics and Asians, the Rejection Rate Is Even Higher." *Washington Post*, May 23, 2018. Accessed October 17, 2021. https://www.washingtonpost.com/realestate/large-numbers-of-loan-applications-get-denied-but-for-blacks-hispanics-and-asians-the-rejection-rate-is-even-higher/2018/05/22/dac19ffc-5d1b-11e8-9ee3-49d6d4814c4c_story.html.

Harwell, Drew. "Wrongfully Arrested Man Sues Detroit Police over False Facial Recognition Match." *Washington Post*, April 13, 2021. Accessed April 13, 2021. https://www.washingtonpost.com/technology/2021/04/13/facial-recognition-false-arrest-lawsuit.

Heaven, Will Douglas. "Predictive Policing Algorithms Are Racist. They Need to Be Dismantled." *MIT Technology Review*, July 17, 2020. Accessed July 30, 2020. https://www.technologyreview.com/2020/07/17/1005396/predictive-policing-algorithms-racist-dismantled-machine-learning-bias-criminal-justice.

Heffron, Anna S., Rohan Khazanchi, Naomi Nkinsi, Joel A. Bervell, Jessica P. Cerdeña, James A. Diao, Leo Gordon Eisenstein, et al. "Trainee Perspectives on Race, Antiracism, and the Path toward Justice in Kidney Care." *Clinical Journal of the American Society of Nephrology*, July 8, 2022. https://doi.org/10.2215/CJN.02500222.

Hess, Abigail Johnson. "Rich Students Get Better SAT Scores—Here's Why." CNBC, October 3, 2019. https://www.cnbc.com/2019/10/03/rich-students-get-better-sat-scores -heres-why.html.

Hicks, Mar. "Hacking the Cis-Tem." *IEEE Annals of the History of Computing* 41, no. 1 (January 1, 2019): 20–33. https://doi.org/10.1109/MAHC.2019.2897667.

Hill, Kashmir. "Another Arrest, and Jail Time, Due to a Bad Facial Recognition Match." *New York Times*, December 29, 2020. https://www.nytimes.com/2020/12 /29/technology/facial-recognition-misidentify-jail.html.

Hill, Kashmir. "Wrongfully Accused by an Algorithm." *New York Times*, June 24, 2020. https://www.nytimes.com/2020/06/24/technology/facial-recognition-arrest.html.

Hill, Michael. "NY Social Service Agency Sued for Not Allowing X Gender Mark." AP News, March 29, 2021. https://apnews.com/article/new-york-lawsuits-social-services -coronavirus-pandemic-medicaid-3e2f1cbae9d2acf2f3994df5039375b8.

Hurtle, Cara Esten. "I'm a Trans Woman. Google Photos Doesn't Know How to Categorize Me." Fast Company, January 24, 2020. https://www.fastcompany.com /90455454/im-a-trans-woman-google-photos-doesnt-know-how-to-categorize-me.

IBM. "AI Governance: Ensuring Your AI Is Transparent, Compliant, and Trustwor- thy." IBM, April 26, 2021. https://www.ibm.com/analytics/common/smartpapers/ai -governance-smartpaper.

Imani. "The Call Is Coming from Inside the House: White Supremacy and the Dis- ability Community." Crutches and Spice, January 27, 2021. https://crutchesandspice .com/2021/01/27/the-call-is-coming-from-inside-the-house-white-supremacy-and -the-disability-community.

Jackson, Kaitlin. "Challenging Facial Recognition Software in Criminal Court." National Association of Criminal Defense Lawyers, 2019. https://www.nacdl .org/getattachment/548c697c-fd8e-4b8d-b4c3-2540336fad94/challenging-facial -recognition-software-in-criminal-court_july-2019.pdf.

Jackson, Liz. "A Community Response to a #DisabilityDongle." *Medium* (blog), April 22, 2019. https://eejackson.medium.com/a-community-response-to-a-disabilitydongle -d0a37703d7c2.

Jackson, Sarah J., Moya Bailey, and Brooke Foucault Welles. *#HashtagActivism: Net- works of Race and Gender Justice.* Cambridge, MA: MIT Press, 2020. https://doi.org/10 .7551/mitpress/10858.001.0001.

Jillson, Elisa. "Aiming for Truth, Fairness, and Equity in Your Company's Use of AI." Federal Trade Commission Business Blog, April 19, 2021. https://www.ftc.gov/news -events/blogs/business-blog/2021/04/aiming-truth-fairness-equity-your-companys -use-ai.

Johnson, David L. "The Role of the Digital Computer in Mechanical Translation of Languages." In *Proceedings of the May 6–8, 1958, Western Joint Computer Conference: Contrasts in Computers*, 161–165. IRE-ACM-AIEE '58 (Western). New York: Association for Computing Machinery, 1958. https://doi.org/10.1145/1457769.1457815.

Kadambi, Achuta. "Achieving Fairness in Medical Devices." *Science* 372, no. 6537 (April 2, 2021): 30. https://doi.org/10.1126/science.abe9195.

Kamara, Seny. "Crypto for the People, Invited Talk at Crypto 2020 by Seny Kamara." YouTube, August 19, 2020. https://www.youtube.com/watch?v=Ygq9ci0GFhA.

Kendi, Ibram X. "'The Difference between Being 'Not Racist' and Antiracist.'" Ted2020, May 2020. Accessed October 24, 2021. https://www.ted.com/talks/ibram_x_kendi_the_difference_between_being_not_racist_and_antiracist/transcript.

Keyes, Os. "The Misgendering Machines: Trans/HCI Implications of Automatic Gender Recognition." *Proceedings of the ACM on Human-Computer Interaction* 2, no. CSCW (November 2018): 1–22. https://doi.org/10.1145/3274357.

Kleinberg, J., S. Mullainathan, and M. Raghavan. "Inherent Trade-Offs in the Fair Determination of Risk Scores." ArXiv E-Prints, September 2016. https://arxiv.org/abs/1609.05807.

Kolata, Gina. "Many Medical Decision Tools Disadvantage Black Patients." *New York Times*, June 17, 2020. https://www.nytimes.com/2020/06/17/health/many-medical-decision-tools-disadvantage-black-patients.html.

Lebovitz, Sarah, Hila Lifshitz-Assaf, and Natalia Levina. "To Engage or Not to Engage with AI for Critical Judgments: How Professionals Deal with Opacity When Using AI for Medical Diagnosis." *Organization Science* 33, no. 1 (January 2022): 126–148. https://doi.org/10.1287/orsc.2021.1549.

Lecher, Colin, and Maddy Varner. "Remote Learning during the Pandemic Has Hit Vulnerable Students the Hardest." The Markup, August 13, 2020. Accessed November 1, 2021. https://themarkup.org/coronavirus/2020/08/13/remote-learning-attendance-inequity-florida-schools.

Lécuyer, Christophe. "From Clean Rooms to Dirty Water: Labor, Semiconductor Firms, and the Struggle over Pollution and Workplace Hazards in Silicon Valley." *Information & Culture* 52, no. 3 (August 1, 2017): 304–333. https://doi.org/10.7560/IC52302.

"Letter to AMS Notices: Boycott Collaboration with Police." Google Docs. Accessed October 10, 2021. https://docs.google.com/forms/d/e/1FAIpQLSfdmQGrgdCBCexTrpne7KXUzpbiI9LeEtd0Am-qRFimpwuv1A/viewform?usp=embed_facebook.

Linder, Courtney. "Why Hundreds of Mathematicians Are Boycotting Predictive Policing." *Popular Mechanics*, July 20, 2020. https://www.popularmechanics.com/science/math/a32957375/mathematicians-boycott-predictive-policing.

Liu, Yuan, Ayush Jain, Clara Eng, David H. Way, Kang Lee, Peggy Bui, Kimberly Kanada, et al. "A Deep Learning System for Differential Diagnosis of Skin Diseases." *Nature Medicine* 26, no. 6 (June 2020): 900–908. https://doi.org/10.1038/s41591-020 -0842-3.

Longley, Robert. "The History of Modern Policing." ThoughtCo. Updated July 13, 2020. Accessed May 26, 2021. https://www.thoughtco.com/the-history-of-modern -policing-974587.

Lorde, Audre. *Sister Outsider: Essays and Speeches*. Berkeley, CA: Crossing Press, 2007.

Lough, Catherine. "IB Results: Anger Grows over Grading 'Scandal.'" Tes, July 8, 2020. https://www.tes.com/news/coronavirus-ib-results-day-2020-controversy-grows -over-grading-scandal.

Mack, Julie. "How Technology and Retail Trends Have Changed Shoplifting, for Better and Worse," mlive, December 8, 2019. https://www.mlive.com/news/2019 /12/how-technology-and-retail-trends-have-changed-shoplifting-for-better-and -worse.html.

Martinez, Emmanuel, and Malena Carollo. "Markup Investigation Cited by Officials in Announcing New Mortgage Discrimination Enforcement." The Markup, October 22, 2021. Accessed July 4, 2022. https://themarkup.org/denied/2021/10/22/markup -investigation-cited-by-officials-in-announcing-new-mortgage-discrimination -enforcement.

Martinez, Emmanuel, and Lauren Kirchner. "The Secret Bias Hidden in Mortgage-Approval Algorithms." The Markup, August 25, 2021. https://themarkup.org/denied /2021/08/25/the-secret-bias-hidden-in-mortgage-approval-algorithms.

Mazmanian, Melissa, and Christine M. Beckman. "'Making' Your Numbers: Engendering Organizational Control through a Ritual of Quantification." *Organization Science* 29, no. 3 (June 1, 2018): 357–379. https://doi.org/10.1287/orsc.2017.1185.

McDonald, Soraya Nadia, McDonald, "'Believe Me' Means Believing That Black Women Are People." In *Believe Me: How Trusting Women Can Change the World*, edited by Jessica Valenti and Jaclyn Friedman, 64–76. New York: Seal Press, 2020.

McGrory, Kathleen, and Neil Bedi. "The Man behind the Machine." *Tampa Bay Times*, December 24, 2020. https://projects.tampabay.com/projects/2020/investigations /police-pasco-sheriff-targeted/chris-nocco.

McGrory, Kathleen, and Neil Bedi. "Targeted." *Tampa Bay Times*, September 3, 2020. https://projects.tampabay.com/projects/2020/investigations/police-pasco-sheriff -targeted/intelligence-led-policing.

McGuinness, Tara Dawson, and Hana Schank. *Power to the Public: The Promise of Public Interest Technology*. Princeton, NJ: Princeton University Press, 2021.

McIlwain, Charlton D. *Black Software: The Internet and Racial Justice, from the AfroNet to Black Lives Matter*. New York: Oxford University Press, 2020.

Melamed, Samantha. "A Leaked Memo Suggests Philly Police Use Vehicle Stops to Get around Stop-and-Frisk Reform." *Philadelphia Inquirer*, March 2, 2021. Accessed March 12, 2021. https://www.inquirer.com/news/philadelphia-police-stop-and-frisk-racial-disparities-vehicle-20210302.html.

Metz, Cade, and Kashmir Hill. "Here's a Way to Learn If Facial Recognition Systems Used Your Photos." *New York Times*, January 31, 2021. https://www.nytimes.com/2021/01/31/technology/facial-recognition-photo-tool.html.

Mukherjee, Siddhartha. "A.I. Versus M.D." *New Yorker*, April 3, 2017. Accessed May 6, 2021. https://www.newyorker.com/magazine/2017/04/03/ai-versus-md.

NASP. "The Shoplifting Problem." *NASP* (blog). Accessed March 4, 2021. https://www.shopliftingprevention.org/the-shoplifting-problem.

National Kidney Foundation. "Establishing a Task Force to Reassess the Inclusion of Race in Diagnosing Kidney Diseases," July 2, 2020. https://www.kidney.org/news/establishing-task-force-to-reassess-inclusion-race-diagnosing-kidney-diseases.

National Kidney Foundation. "Race and EGFR: What Is the Controversy?," August 3, 2020. https://www.kidney.org/atoz/content/race-and-egfr-what-controversy.

National Kidney Foundation. "Understanding African American and Non-African American eGFR Laboratory Results," https://www.kidney.org/atoz/content/race-and-egfr-what-controversy.

New York City Department of Education. "School Quality Reports: Using 'Comparison Group' Results to Better Understand a School's Performance," December 4, 2017. https://infohub.nyced.org/docs/default-source/default-document-library/schoolqualityreports_comparisongroupdescription_20171204fc489b8b347e46bf801e38ceed4b3069.pdf?sfvrsn=e747dd75_2.

New York Civil Liberties Union. "Stop-and-Frisk Data," January 2, 2012. https://www.nyclu.org/en/stop-and-frisk-data.

Ng, Alfred. "Police Say They Can Use Facial Recognition, Despite Bans." The Markup. Accessed January 28, 2021. https://themarkup.org/news/2021/01/28/police-say-they-can-use-facial-recognition-despite-bans.

Ng, Alfred. "Senator Calls Facebook Response on Discriminatory Ads 'Inadequate.'" The Markup, May 24, 2021. Accessed October 24, 2021. https://themarkup.org/citizen-browser/2021/05/24/senator-calls-facebook-response-on-discriminatory-ads-inadequate.

Nix, Naomi. "Facebook Sees Slight Decline in Female Worker Representation." *Bloomberg.com*, July 15, 2021. https://www.bloomberg.com/news/articles/2021-07-15/facebook-fb-female-worker-representation-declines.

Noble, Safiya Umoja. *Algorithms of Oppression: How Search Engines Reinforce Racism.* New York: New York University Press, 2018.

Noble, Safiya. "The Loss of Public Goods to Big Tech." *Noēma*, July 1, 2020. Accessed September 7, 2020. https://www.noemamag.com/the-loss-of-public-goods-to-big-tech.

NYU Communications. "Gender Identity in Albert." Accessed July 3, 2022. http://www.nyu.edu/content/nyu/en/students/student-information-and-resources/registration-records-and-graduation/forms-policies-procedures/gender-identity.

O'Neil, Cathy. *Weapons of Math Destruction: How Big Data Increases Inequality and Threatens Democracy.* New York: Crown Publishers, 2016.

"Passport Facial Recognition Checks Fail to Work with Dark Skin," BBC News, October 9, 2019. https://www.bbc.com/news/technology-49993647.

Petty, Tawana. Twitter post, June 2, 2021, 2:35 pm. https://twitter.com/Combsthepoet/status/1400159136072048640.

"Predictive Policing." Upturn. Accessed June 3, 2021. https://teamupturn.gitbooks.io/predictive-policing/content.

"ProctorU to Discontinue Exam Integrity Services That Rely Exclusively on AI." ProctorU, May 24, 2021. https://www.proctoru.com/industry-news-and-notes/proctoru-to-discontinue-exam-integrity-services-that-rely-exclusively-on-ai.

Pulitzer Center for Health Journalism. "Pulitzer Spotlight: Policing in the Era of Big Data." Accessed March 7, 2022. https://centerforhealthjournalism.org/content/pulitzer-spotlight-policing-era-big-data.

Quinn, Gerard, Theresia Degener, Anna Bruce, Christine Burke, Joshua Castellino, Padraic Kenna, Ursula Kilkelly, and Shivaun Quinlivan. "Human Rights and Disability: The Current Use and Future Potential of United Nations Human Rights Instruments in the Context of Disability." New York: United Nations, 2002.

Raji, Deborah. Twitter post, June 28, 2021, 10:12 pm. https://twitter.com/rajiinio/status/1409333845795622912.

Raji, Inioluwa Deborah, Morgan Klaus Scheuerman, and Razvan Amironesei. "You Can't Sit with Us: Exclusionary Pedagogy in AI Ethics Eduction." *FAccT '21: Proceedings of the 2021 ACM Conference on Fairness, Accountability, and Transparency.* New York: ACM, 2021. https://doi.org/10.1145/3442188.3445914.

Ray, Rashawn. "How Can We Enhance Police Accountability in the United States?" *Brookings*, August 25, 2020. https://www.brookings.edu/policy2020/votervital/how -can-we-enhance-police-accountability-in-the-united-states.

Rex, Gotham. "N.Y.P.D. Will Stop Using Robot Dog after Backlash." *New York Times*, April 28, 2021. Accessed April 29, 2021. https://www.nytimes.com/2021/04/28 /nyregion/nypd-robot-dog-backlash.html?smid=tw-nytimes&smtyp=cur.

Rhinesmith, Vanessa. "Joint Civil Society Statement—Interventions to Mitigate the Racially Discriminatory Impacts of Emerging Tech." UCLA Center for Critical Internet Inquiry, July 15, 2020. Accessed August 31, 2021. https://www.c2i2.ucla .edu/2020/07/15/joint-civil-society-statement-interventions-to-mitigate-the-racially -discriminatory-impacts-of-emerging-tech-including-ai.

Richardson, Lisa C. "Patterns and Trends in Age-Specific Black-White Differences in Breast Cancer Incidence and Mortality—United States, 1999–2014." *MMWR. Morbidity and Mortality Weekly Report* 65 (2016). https://doi.org/10.15585/mmwr .mm6540a1.

Richardson, Rashida. "Racial Segregation and the Data-Driven Society: How Our Failure to Reckon with Root Causes Perpetuates Separate and Unequal Realities." SSRN Scholarly Paper. Rochester, NY: Social Science Research Network, May 20, 2021. https://papers.ssrn.com/abstract=3850317.

Robbins, Rebecca, and Erin Brodwin. "Patients Aren't Being Told about the AI Systems Advising Their Care." STAT, July 15, 2020. https://www.statnews.com/2020/07/15 /artificial-intelligence-patient-consent-hospitals.

Roberts, Dorothy. "Race Correction." Twitter post, April 30, 2021, 8:26 am. https:// twitter.com/DorothyERoberts/status/1388107500348641283.

Roberts, Sarah T. *Behind the Screen: Content Moderation in the Shadows of Social Media.* New Haven, CT: Yale University Press, 2019.

Ross, Casey. "Machine Learning Is Booming in Medicine. It's Also Facing a Credibility Crisis," STAT, June 2, 2021. https://www.statnews.com/2021/06/02/machine -learning-ai-methodology-research-flaws.

Roy, Elise. "When We Design for Disability, We All Benefit." TedXMidAtlantic, September 2015. https://www.ted.com/talks/elise_roy_when_we_design_for_disability _we_all_benefit.

Ryan-Mosley, Tate, and Jennifer Strong. "The Activist Dismantling Racist Police Algorithms." MIT Technology Review, June 5, 2020. Accessed August 2, 2021. https://www.technologyreview.com/2020/06/05/1002709/the-activist-dismantling -racist-police-algorithms.

Salam, Maya. "For Serena Williams, Childbirth Was a Harrowing Ordeal. She's Not Alone." *New York Times*, January 11, 2018. https://www.nytimes.com/2018/01/11/sports/tennis/serena-williams-baby-vogue.html.

Sankaranarayanan, R. "Screening for Cancer in Low- and Middle-Income Countries." *Annals of Global Health* 80, no. 5 (October 2014): 412–417. https://doi.org/10.1016/j.aogh.2014.09.014.

Scannell, R. Joshua. "Broken Windows, Broken Code." Real Life, August 29, 2016. https://reallifemag.com/broken-windows-broken-code.

Schiffer, Zoe. "Apple Employees Are Organizing, Now under the Banner #Apple-Too." The Verge, August 23, 2021. https://www.theverge.com/2021/8/23/22638150/apple-appletoo-employee-harassment-discord.

Sciacca, Annie. "Oakland Police Give FBI 'Unfettered Access' to License Plate Reader Data, According to Lawsuit." *East Bay Times*, September 4, 2021. https://www.eastbaytimes.com/2021/09/04/oakland-police-give-fbi-unfettered-access-to-license-plate-reader-data-according-to-lawsuit.

Shapiro, Nina. "When the World Went Remote, Communities on the Wrong Side of the Digital Divide Got Shut Out." Center for Health Journalism, July 14, 2021. https://centerforhealthjournalism.org/2021/07/13/when-world-went-remote-communities-wrong-side-digital-divide-got-shut-out.

Singer, Natasha, and Cade Metz. "Many Facial-Recognition Systems Are Biased, Says U.S. Study." *New York Times*, December 19, 2019. https://www.nytimes.com/2019/12/19/technology/facial-recognition-bias.html.

Sjoding, Michael W., Robert P. Dickson, Theodore J. Iwashyna, Steven E. Gay, and Thomas S. Valley. "Racial Bias in Pulse Oximetry Measurement." *New England Journal of Medicine* 383, no. 25 (December 17, 2020): 2477–2478. https://doi.org/10.1056/NEJMc2029240.

Skloot, Rebecca. *The Immortal Life of Henrietta Lacks*. New York: Broadway Paperbacks, 2011.

Sloan, Karen. "Plaintiffs' Firm Eyes Class Action over Bar Exam Tech Problems." Reuters, August 2, 2021. Accessed November 1, 2021. https://www.reuters.com/legal/litigation/plaintiffs-firm-eyes-class-action-over-bar-exam-tech-problems-2021-08-02.

Spiezio, Caroline. "Michigan Software Crash Roils First Online U.S. Bar Exam." *Reuters*, July 28, 2020. https://www.reuters.com/article/lawyer-coronavirus-michigan-idUSL2N2EZ26A.

Stevens, Nikki, Anna Lauren Hoffmann, and Sarah Florini. "The Unremarked Optimum: Whiteness, Optimization, and Control in the Database Revolution."

Review of Communication 21, no. 2 (April 3, 2021): 113–128. https://doi.org/10.1080/15358593.2021.1934521.

Strickland, Eliza. "Racial Bias Found in Algorithms That Determine Health Care for Millions of Patients." *IEEE Spectrum*, October 24, 2019. Accessed April 8, 2021. https://spectrum.ieee.org/the-human-os/biomedical/ethics/racial-bias-found-in-algorithms-that-determine-health-care-for-millions-of-patients.

Stroud, Matt. "An Automated Policing Program Got This Man Shot Twice." The Verge, May 24, 2021. https://www.theverge.com/22444020/heat-listed-csk-entry.

Tang, Hazel. "A New Study Found Many Clinical Algorithms Are Still Subjected to Racial Biases." AIMed, June 22, 2020. https://ai-med.io/clinicians/a-new-study-found-many-clinical-algorithms-are-still-subjected-to-racial-biases.

Tate, Emily. "Why Aren't Schools Using the Apps They Pay For?" EdSurge, November 8, 2018. https://www.edsurge.com/news/2018-11-08-why-aren-t-schools-using-the-apps-they-pay-for.

Thomas, Rachel. "Medicine's Machine Learning Problem." *Boston Review*, October 19, 2020. https://bostonreview.net/science-nature/rachel-thomas-medicines-machine-learning-problem.

Treisman, Rachel. "Bumble Blunder: Man Allegedly Boasts about Capitol Riot on Dating App, Is Arrested." *NPR*, April 23, 2021. https://www.npr.org/2021/04/23/990218018/bumble-blunder-man-allegedly-boasts-about-capitol-riot-on-dating-app-is-arrested.

"U of T Neural Networks Start-Up Acquired by Google." University of Toronto media release, March 12, 2013. https://media.utoronto.ca/media-releases/u-of-t-neural-networks-start-up-acquired-by-google.

Urban, Noah, Jacob Yesh-Brochstein, Erica Raleigh, and Tawana Petty. "A Critical Summary of Detroit's Project Green Light and Its Greater Context." Detroit Community Technology Project, June 9, 2019. https://detroitcommunitytech.org/system/tdf/librarypdfs/DCTP_PGL_Report.pdf?file=1&type=node&id=77&force=.

US Digital Service. "USDS Alumni Network: Hana Schank." *US Digital Service* (blog), July 23, 2020. https://medium.com/the-u-s-digital-service/usds-alumni-network-hana-schank-aecd2d841879.

US General Services Administration. "Universal Design: What Is It?" Section508.gov. Accessed August 21, 2021. https://section508.gov/blog/Universal-Design-What-is-it.

Valenti, Jessica, and Jaclyn Friedman. *Believe Me: How Trusting Women Can Change the World*. New York: Seal Press, 2020.

Verma, Sahil, and Julia Rubin. "Fairness Definitions Explained." In *Proceedings of the International Workshop on Software Fairness*, 1–7. Gothenburg: ACM, 2018. https://doi.org/10.1145/3194770.3194776.

Villarosa, Linda. *Under the Skin: Racism, Inequality, and the Health of a Nation*. New York: Doubleday, 2022.

Waldron, Lucas, and Brenda Medina. "TSA's Body Scanners Are Gender Binary. Humans Are Not." *ProPublica*, August 26, 2019. https://www.propublica.org/article/tsa-transgender-travelers-scanners-invasive-searches-often-wait-on-the-other-side?token=PrSqc58cn7gW-eRA_isLN03ygBJN8E84.

Wallis, Claudia. "Fixing Medical Devices That Are Biased against Race or Gender." *Scientific American*, June 1, 2021. Accessed April 14, 2022. https://www.scientificamerican.com/article/fixing-medical-devices-that-are-biased-against-race-or-gender.

Wall Street Journal. "The Facebook Files," October 1, 2021. https://www.wsj.com/articles/the-facebook-files-11631713039.

Ward, Stephanie Francis, and Lyle Moran. "Thousands of California Bar Exam Takers Have Video Files Flagged for Review." *ABA Journal*, December 8, 2020. https://www.abajournal.com/web/article/thousands-of-california-bar-exam-takers-have-video-files-flagged-for-review.

Warzel, Charlie, and Stuart A. Thompson. "They Stormed the Capitol. Their Apps Tracked Them." Editorial. *New York Times*, February 5, 2021. https://www.nytimes.com/2021/02/05/opinion/capitol-attack-cellphone-data.html.

"What Is Color Blindness?" American Academy of Ophthalmology, April 6, 2021. https://www.aao.org/eye-health/diseases/what-is-color-blindness.

Whittaker, Meredith, Meryl Alper, Olin College, Liz Kaziunas, and Meredith Ringel Morris. "Disability, Bias, and AI," n.d. https://ainowinstitute.org/disabilitybiasai-2019.pdf.

Williams, Robert. "I Was Wrongfully Arrested Because of Facial Recognition. Why Are Police Allowed to Use It?" Editorial. *Washington Post*, June 24, 2020. Accessed July 20, 2021. https://www.washingtonpost.com/opinions/2020/06/24/i-was-wrongfully-arrested-because-facial-recognition-why-are-police-allowed-use-this-technology.

Wilson, Christo, Avijit Ghosh, Shan Jiang, Alan Mislove, Lewis Baker, Janelle Szary, Kelly Trindel, and Frida Polli. "Building and Auditing Fair Algorithms: A Case Study in Candidate Screening." In *Proceedings of the 2021 ACM Conference on Fairness, Accountability, and Transparency*, 666–677. Virtual Event Canada: ACM, 2021. https://doi.org/10.1145/3442188.3445928.

Wu, Eric, Kevin Wu, Roxana Daneshjou, David Ouyang, Daniel E. Ho, and James Zou. "How Medical AI Devices Are Evaluated: Limitations and Recommendations from an

Analysis of FDA Approvals." *Nature Medicine* 27, no. 4 (April 2021): 582–584. https://doi.org/10.1038/s41591-021-01312-x.

Wu, Nan, Jason Phang, Jungkyu Park, Yiqiu Shen, S. Gene Kim, Laura Heacock, Linda Moy, Kyunghyun Cho, and Krzysztof J. Geras. "The NYU Breast Cancer Screening Dataset v1.0," September 16, 2021. https://cs.nyu.edu/~kgeras/reports/datav1.0.pdf.

"Wyden, Booker and Clarke Introduce Algorithmic Accountability Act of 2022 to Require New Transparency and Accountability for Automated Decision Systems." Website of U.S. Senator Ron Wyden of Oregon. Accessed March 8, 2022. https://www.wyden.senate.gov/news/press-releases/wyden-booker-and-clarke-introduce-algorithmic-accountability-act-of-2022-to-require-new-transparency-and-accountability-for-automated-decision-systems.

Young, Damon. "The Least Livable Body in America's Most Livable City." *Esquire*, October 22, 2020. https://www.esquire.com/news-politics/a34315507/racism-public-health-crisis-black-women-pittsburgh.

Young, Damon. "Racism Makes Me Question Everything. I Got the Vaccine Anyway." Editorial. *New York Times*, April 9, 2021. https://www.nytimes.com/2021/04/09/opinion/racism-covid-vaccine.html.

Zakrzewski, Cat, Gerrit De Vynck, Niha Masih, and Shibani Mahtani. "How Facebook Neglected the Rest of the World, Fueling Hate Speech and Violence in India," *Washington Post*, October 24, 2021. https://www.washingtonpost.com/technology/2021/10/24/india-facebook-misinformation-hate-speech.

Zaman, Samihah. "IB to Make Adjustments to Results Awarded for May 2020 Following Multiple Review Requests." Gulf News, August 17, 2020. https://gulfnews.com/uae/education/ib-to-make-adjustments-to-results-awarded-for-may-2020-following-multiple-review-requests-1.73241291.

Zuberi, Tukufu. *Thicker Than Blood: How Racial Statistics Lie*. Minneapolis: University of Minnesota Press, 2001.

Zuckerman, Ethan. "To the Future Occupants of My Office at the MIT Media Lab," August 15, 2020. https://ethanzuckerman.com/2020/08/15/to-the-future-occupants-of-my-office-at-the-mit-media-lab.

Index